BUSINESS ISSUES, COMPETITION AND ENTREPRENEURSHIP

SMALL BUSINESS INNOVATION RESEARCH AND SMALL BUSINESS TECHNOLOGY TRANSFER PROGRAMS

BACKGROUND AND ISSUES

BUSINESS ISSUES, COMPETITION AND ENTREPRENEURSHIP

Additional books in this series can be found on Nova's website under the Series tab.

Additional e-books in this series can be found on Nova's website under the e-book tab.

BUSINESS ISSUES, COMPETITION AND ENTREPRENEURSHIP

SMALL BUSINESS INNOVATION RESEARCH AND SMALL BUSINESS TECHNOLOGY TRANSFER PROGRAMS

BACKGROUND AND ISSUES

THOMAS H. CHAVEZ
EDITOR

nova publishers
New York

Library of Congress Cataloging-in-Publication Data

ISBN: 978-1-63482-532-0

Published by Nova Science Publishers, Inc. † New York

CONTENTS

PREFACE

In establishing the program, Congress found that technological innovation plays an important role in job creation, productivity improvements, and U.S. competitiveness; that small businesses are among the most cost-effective performers of R&D and particularly capable of bringing R&D results to market in the form of new products; and that despite the role of small businesses as — the principal source of significant innovations in the Nation, the vast majority of federally funded R&D is performed by large businesses, universities, and federal laboratories. This book provides background and examines issues in the SBIR and small business technology transfer programs.

Chapter 1 - The Small Business Innovation Research (SBIR) program was established in 1982 by the Small Business Innovation Development Act (P.L. 97-219) to increase the participation of small innovative companies in federally funded R&D. The act requires federal agencies with extramural R&D budgets of $100 million or more to set aside a portion of these funds to finance an agency-run SBIR program. As of 2014, 11 federal agencies operate SBIR programs. A complementary program, the Small Business Technology Transfer (STTR) program, was created by the Small Business Research and Development Enhancement Act of 1992 (P.L. 102-564) to facilitate the commercialization of university and federal R&D by small companies. Agencies with extramural R&D budgets of $1 billion or more are required to set aside a portion of these funds to finance an agency-run STTR program. As of 2014, five federal agencies operate STTR programs.

Both the SBIR and STTR programs have three phases. Phase I funds feasibility-related research and development (R&D) related to agency requirements. Phase II supports further R&D efforts initiated in Phase I that meet particular program needs and that exhibit potential for commercial application. Phase III is focused on commercialization of the results of Phase I and Phase II grants, however the SBIR and STTR programs do not provide funding in Phase III.

The SBIR and STTR programs have been extended and reauthorized several times since their initial enactments. Most recently, the programs were reauthorized through September 30, 2017 under the SBIR/STTR Reauthorization Act of 2011 which was enacted as Division E of the National Defense Authorization Act for Fiscal Year 2012 (P.L. 112-81). Among its provisions, the act incrementally increases the set-aside for the SBIR effort to 3.2% by FY2017 and beyond; incrementally expands the set-aside for the STTR activity to 0.45% in FY2016 and beyond; increases the amount of Phase I and Phase II awards; allows recipients of a Phase I award from one federal agency to apply for a Phase II award from another agency to pursue the original work; allows the National Institutes of Health, the Department of Energy, and the National Science Foundation to award up to 25% of SBIR funds to small businesses that are majority-owned by venture capital companies, hedge funds, or private equity firms, and allows other agencies to award up to 15% of SBIR funds to such firms; creates commercialization pilot programs; and expands oversight activities, among other things.

Through FY2011, federal agencies had made more than 133,000 awards totaling $33.7 billion under the SBIR and STTR programs. In FY2011, agencies awarded $2.224 billion in SBIR funding. The Department of Defense (DOD) and Department of Health and Human Services (HHS) accounted for more than three-fourths of SBIR funding in FY2011. While more than two-thirds of SBIR grants made in FY2011 were Phase I awards, more than three-fourths of SBIR funding went to Phase II awards. In FY2011, agencies awarded $251.2 million in STTR funding. DOD and HHS accounted for nearly four-fifths of STTR funding. Like the SBIR program, most STTR grants (76%) were for Phase I awards, while most funding (76%) went to Phase II awards.

In exercising its oversight authorities for the SBIR and STTR programs, Congress has placed a strong emphasis on monitoring the implementation and effects of changes made by the 2011 reauthorization act. In particular, Congress has expressed continuing interest in the participation of majority-owned venture capital firms in the SBIR program, the effectiveness of efforts seeking to improve commercialization outcomes, the share of awards and funding received by women-owned and minority and disadvantaged firms, and the SBA's agency coordination, policy guidance, data collection, and dissemination responsibilities.

Chapter 2 - Federal agencies have awarded more than 156,000 contracts and grants, totaling nearly $40 billion, through the SBIR and STTR programs to small businesses to develop and commercialize innovative technologies. The Small Business Act requires agencies with extramural R&D budgets that meet certain thresholds for participation—$100 million for SBIR and $1 billion for STTR—must spend a percentage of these annual budgets on the programs. The agencies are to report on their activities to SBA and, in turn, SBA is to report to Congress.

The 2011 reauthorization of the programs mandated GAO to review compliance with spending and reporting requirements, and other program aspects. This report addresses, for fiscal year 2012, (1) the extent to which agencies complied with spending requirements, (2) the extent to which agencies and SBA complied with certain reporting requirements, (3) the potential effects of basing spending requirements on agencies' total R&D budgets, and (4) what is known about the amounts spent administering the programs. GAO reviewed agency spending data and required reports for fiscal year 2012 and interviewed program officials from SBA and the participating agencies.

Chapter 3 - Starting in 2013, implementation of the National Defense Authorization Act for Fiscal Year 2012 (NDAA) enabled agencies to allow small businesses that are majority owned by multiple venture capital operating companies, hedge funds, or private equity firms (majority-owned portfolio companies) to participate in the Small Business Innovation Research (SBIR) program. In 1982, Congress established the SBIR program to, among other things, stimulate technological innovation related to federal agency goals and missions, use small businesses to meet federal research or research and development (R&D) needs, and increase private sector commercialization of innovations derived from federal R&D. In fiscal year 2013, federal agencies made more than 4,000 awards totaling more than $2.1 billion through the SBIR program. Over the program's first two decades, some majority-owned portfolio companies participated in SBIR, without any apparent adverse consequence for the program's operation and achievements, according to a study by the National Academy of Sciences. However, in 2003, a Small Business Administration (SBA) administrative law judge issued a decision holding that such companies were ineligible to receive SBIR awards. The decision raised concerns that some portfolio companies were being unfairly excluded from SBIR and that important research was not being funded. Others supported the decision, contending that small businesses without venture capital funding would not be able to compete against small businesses with such funding.

Chapter 4 - Report of a SBIR/STTR Interagency Policy Committee on SBIR/STTR Outreach.

Chapter 5 - Report of a SBIR/STTR Interagency Policy Committee on SBIR/STTR Commercialization.

Chapter 6 - Report of a SBIR/STTR Interagency Policy Committee on SBIR/STTR Standard Evaluation Framework.

In: Small Business Innovation Research … ISBN: 978-1-63482-532-0
Editor: Thomas H. Chavez

Chapter 1

SMALL BUSINESS INNOVATION RESEARCH AND SMALL BUSINESS TECHNOLOGY TRANSFER PROGRAMS*

John F. Sargent Jr.

SUMMARY

The Small Business Innovation Research (SBIR) program was established in 1982 by the Small Business Innovation Development Act (P.L. 97-219) to increase the participation of small innovative companies in federally funded R&D. The act requires federal agencies with extramural R&D budgets of $100 million or more to set aside a portion of these funds to finance an agency-run SBIR program. As of 2014, 11 federal agencies operate SBIR programs. A complementary program, the Small Business Technology Transfer (STTR) program, was created by the Small Business Research and Development Enhancement Act of 1992 (P.L. 102-564) to facilitate the commercialization of university and federal R&D by small companies. Agencies with extramural R&D budgets of $1 billion or more are required to set aside a portion of these funds to finance an agency-run STTR program. As of 2014, five federal agencies operate STTR programs.

* This is an edited, reformatted and augmented version of a Congressional Research Service publication R43695, prepared for Members and Committees of Congress, dated August 26, 2014.

Both the SBIR and STTR programs have three phases. Phase I funds feasibility-related research and development (R&D) related to agency requirements. Phase II supports further R&D efforts initiated in Phase I that meet particular program needs and that exhibit potential for commercial application. Phase III is focused on commercialization of the results of Phase I and Phase II grants, however the SBIR and STTR programs do not provide funding in Phase III.

The SBIR and STTR programs have been extended and reauthorized several times since their initial enactments. Most recently, the programs were reauthorized through September 30, 2017 under the SBIR/STTR Reauthorization Act of 2011 which was enacted as Division E of the National Defense Authorization Act for Fiscal Year 2012 (P.L. 112-81). Among its provisions, the act incrementally increases the set-aside for the SBIR effort to 3.2% by FY2017 and beyond; incrementally expands the set-aside for the STTR activity to 0.45% in FY2016 and beyond; increases the amount of Phase I and Phase II awards; allows recipients of a Phase I award from one federal agency to apply for a Phase II award from another agency to pursue the original work; allows the National Institutes of Health, the Department of Energy, and the National Science Foundation to award up to 25% of SBIR funds to small businesses that are majority-owned by venture capital companies, hedge funds, or private equity firms, and allows other agencies to award up to 15% of SBIR funds to such firms; creates commercialization pilot programs; and expands oversight activities, among other things.

Through FY2011, federal agencies had made more than 133,000 awards totaling $33.7 billion under the SBIR and STTR programs. In FY2011, agencies awarded $2.224 billion in SBIR funding. The Department of Defense (DOD) and Department of Health and Human Services (HHS) accounted for more than three-fourths of SBIR funding in FY2011. While more than two-thirds of SBIR grants made in FY2011 were Phase I awards, more than three-fourths of SBIR funding went to Phase II awards. In FY2011, agencies awarded $251.2 million in STTR funding. DOD and HHS accounted for nearly four-fifths of STTR funding. Like the SBIR program, most STTR grants (76%) were for Phase I awards, while most funding (76%) went to Phase II awards.

In exercising its oversight authorities for the SBIR and STTR programs, Congress has placed a strong emphasis on monitoring the implementation and effects of changes made by the 2011 reauthorization act. In particular, Congress has expressed continuing interest in the participation of majority-owned venture capital firms in the SBIR program, the effectiveness of efforts seeking to improve commercialization outcomes, the share of awards and funding received by women-owned and minority and disadvantaged firms, and the SBA's agency coordination, policy guidance, data collection, and dissemination responsibilities.

OVERVIEW

The Small Business Innovation Research (SBIR) program was established by Congress in 1982 to expand the role of small businesses in federal research and development (R&D). In establishing the program, Congress found that technological innovation plays an important role in job creation, productivity improvements, and U.S. competitiveness; that small businesses are among the most cost-effective performers of R&D and particularly capable of bringing R&D results to market in the form of new products; and that despite the role of small businesses as "the principal source of significant innovations in the Nation," the vast majority of federally funded R&D is performed by large businesses, universities, and federal laboratories.[1] With this in mind, Congress established the SBIR program to advance four objectives:

- to stimulate innovation,
- to use small businesses to meet federal R&D needs,
- to foster and encourage the participation of minority and disadvantaged persons in technological innovation, and
- to increase private sector commercialization of innovations derived from federally-funded R&D.[2]

In 1992, Congress established the Small Business Technology Transfer (STTR) program.[3] Similar in design to the SBIR program, STTR was created to facilitate the commercialization of university and federal R&D by small companies.

The SBIR and STTR programs have been reauthorized on multiple occasions, most recently by the SBIR/STTR Reauthorization Act of 2011 (P.L. 112-81), which authorizes both programs through FY2017.[4] Highlights of this law are provided in "SBIR/STTR Reauthorization Act of 2011" later in this report.

Execution of the SBIR and STTR programs is decentralized. Both the SBIR and STTR statutes require that federal agencies with extramural R&D budgets in excess of specified amounts set aside a percentage of such funds to conduct their own SBIR and STTR programs.[5] Currently, 11 federal departments and agencies operate SBIR programs and 5 operate STTR programs. The Small Business Administration (SBA) helps to coordinate the SBIR and STTR programs, establishes overall policy guidance, reviews agencies' progress, and reports annually to Congress on the operation of the programs.

Through FY2011, federal agencies had made more than 133,000 SBIR and STTR awards to small businesses to develop and commercialize innovative technologies. The total amount awarded was $33.7 billion. *Figure 1* shows SBIR and STTR funding for FY2000-FY2011.

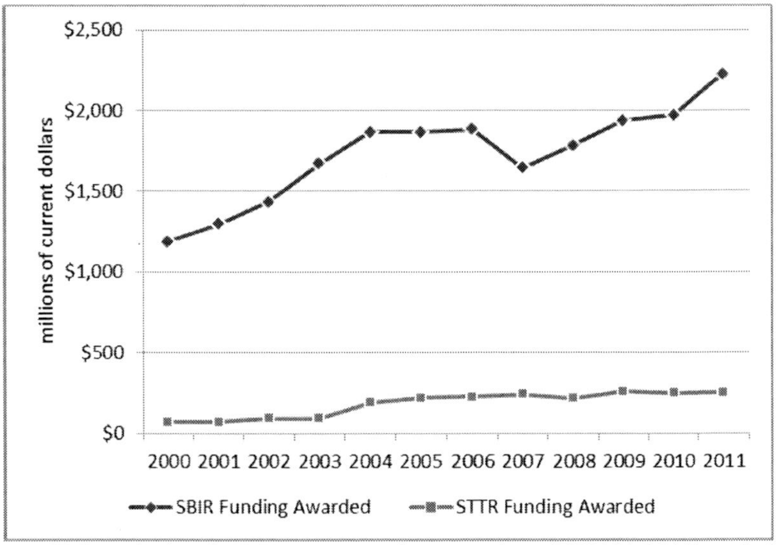

Source: CRS analysis of SBIR.gov annual report data (FY2000-2010) and Small Business Innovation Research (SBIR) and Small Business Technology Transfer (STTR) Program Annual Report FY2009-FY2011 (FY2011).

Figure 1. SBIR and STTR Funding, FY2000-FY2011. (Total of Phase I and Phase II Awards for SBIR and STTR programs).

This report provides information on the legislative foundations, structure, operation, and current and historical funding levels of the SBIR and STTR programs; summarizes the most recent legislative changes to the programs; provides highlights of external reviews of the program; and identifies and discusses selected policy issues.

DATA SOURCES AND LIMITATIONS

The SBA, through its SBIR.gov website, makes available certain data on SBIR and STTR awards through FY2013 (and some awards data for FY2014).

However, the SBA has communicated to CRS that the data for FY2012 and later years have not been cleared by the Office of Management and Budget. For this report, CRS has relied on the SBIR.gov website for data from the inception of the SBIR and STTR programs through FY2010. For FY2011, this report relies on data provided directly to CRS by the SBA.[6] However, the FY2011 data set has some inconsistencies; in several cases, the sum of individual agency amounts does not correspond to the stated total. The SBA informed CRS that some agency numbers were modified prior to publication of the FY2011 figures, but that the totals for each row were inadvertently not recalculated.[7] Accordingly, for this report CRS uses the sum of the individual agency amounts rather than the totals provided by the SBA.

SMALL BUSINESS INNOVATION RESEARCH

SBIR Overview

The Small Business Innovation Research program was established under the Small Business Innovation Development Act of 1982 (P.L. 97-219) and subsequently reauthorized or extended multiple times, most recently in 2011 when the program was reauthorized through September 30, 2017.[8] Under the program, each federal agency with an extramural R&D budget greater than $100 million is required to allocate a portion of that funding to conduct a multi-phase R&D grant program for small businesses. The objectives of the SBIR program include stimulating technological innovation; increasing the use of the small business community to meet federal R&D needs; fostering and encouraging participation in innovation and entrepreneurship by socially and economically disadvantaged individuals; and expanding private-sector commercialization of innovations resulting from federally funded R&D.

Currently, 11 federal agencies participate in the SBIR program: the Departments of Agriculture (USDA), Commerce (DOC), Defense (DOD), Education (ED), Energy (DOE), Health and Human Services (HHS), Homeland Security (DHS), and Transportation (DOT); the Environmental Protection Agency (EPA); the National Aeronautics and Space Administration (NASA); and the National Science Foundation (NSF).

Under the 2011 reauthorization, the minimum percentage of extramural R&D funds that agencies are required to set aside for the SBIR program increases 0.1% per year for five years, from 2.5% in FY2011 to 3.0% in FY2016, then increases to 3.2% for FY2017 and each fiscal year thereafter.

Agencies may opt to exceed these minimum percentages. In FY2011, the aggregate level of SBIR funding for all federal agencies was $2.119 billion, approximately 2.6% of the participating agencies' aggregate extramural R&D funding. However, a recent report by the Government Accountability Office (GAO) found that some agencies did not comply with the SBIR and STTR spending requirements.[9] This issue is addressed in greater detail in "Agency Compliance with Mandatory Minimum Expenditure Levels."

Each participating agency operates its own SBIR program under the provisions of the law and regulations, as well as with the policy directive issued by the U.S. Small Business Administration (SBA) in its *Small Business Innovation Research Program Policy Directive* (referred to hereinafter as the *SBIR Program Policy Directive*).[10] According to some analysts, this approach allows for general consistency across SBIR programs, while allowing each agency a substantial degree of control and flexibility in the execution of its program in alignment with its overall mission and priorities.[11] (See "Improving Technology Commercialization, Trade-Offs Among Program Objectives" for related discussion.)

SBIR Phases

The SBIR program is a three-phase program. The purposes and parameters of each phase are discussed below.

Phase I

In Phase I, an agency solicits contract proposals or grant applications to conduct feasibility-related experimental or theoretical research or research and development (R/R&D) related to agency requirements. The scope of the topic(s) in the solicitation may be broad or narrow, depending on the needs of the agency. Phase I grants are intended to determine "the scientific and technical merit and feasibility of ideas that appear to have commercial potential."[12] Generally, SBIR Phase I awards are not to exceed $150,000, though the law provides agencies with the authority to issue awards that exceed this amount (the Phase I award guideline) by as much as 50%. In addition, agencies may request a waiver from the SBA to exceed the award guideline by more than 50% for a specific topic. In general, the period of performance for Phase I awards is up to six months, though agencies may allow for a longer performance period for a particular project.

Phase II

Phase II grants are intended to further R/R&D efforts initiated in Phase I that meet particular program needs and that exhibit potential for commercial application. In general, only Phase I grant recipients are eligible for Phase II grants. There are two exceptions to this guideline: (1) a federal agency may issue an SBIR Phase II award to a Small Business Technology Transfer (STTR) Phase I awardee to further develop the work performed under the STTR Phase I award;[13] and (2) through FY2017, the National Institutes of Health (NIH), DOD, and ED are authorized to make Phase II grants to small businesses that did not receive Phase I awards. Exercise of this authority requires a written determination from the agency head that the small business has demonstrated the scientific and technical merit and feasibility of the ideas and that the ideas appear to have commercial potential.[14]

Phase II awards are to be based on the results achieved in Phase I (when applicable) and the scientific and technical merit and commercial potential of the project proposed in Phase II as evidenced by: the small business concern's record of successfully commercializing SBIR or other research; the existence of second phase funding commitments from private sector or non-SBIR funding sources; the existence of third phase, follow-on commitments for the subject of the research; and the presence of other indicators of the commercial potential of the idea.[15]

The *SBIR Program Policy Directive* generally limits SBIR Phase II awards to $1 million (the Phase II award guideline), though the directive provides agencies with the authority to issue an award that exceeds this amount by as much as 50% (for an amount up to $1.5 million). As with Phase I grants, agencies may request a waiver from the SBA to exceed the Phase II award guideline by more than 50% for a specific topic. In general, the period of performance for Phase II awards is not to exceed two years, though agencies may allow for a longer performance period for a particular project. Agencies may make a sequential Phase II award to continue the work of an initial Phase II award. This sequential Phase II award is also subject to the $1 million Phase II guideline and agencies' authority to exceed the guideline by up to 50%. Thus, agencies may award up to $3 million in Phase II awards for a particular project to a single recipient at the agency's discretion, and potentially more if the agency requests and receives a waiver from the SBA. For sequential Phase II awards, some agencies require third party matching of the SBIR funds.

Phase III

Phase III of the SBIR program is focused on the commercialization of the results achieved with Phase I and Phase II SBIR funding. The SBIR program *does not* provide funding in Phase III. Phase III funding is expected, generally, to be generated in the private sector. However, some agencies may use non-SBIR funds for Phase III funding to support additional R&D or contracts for products, processes, or services intended for use by the federal government. In addition, the 2011 reauthorization act directs agencies and prime contractors "to the greatest extent practicable," to facilitate the commercialization of SBIR and STTR through the use of Phase III awards, including sole source awards.[16]

Technical Assistance

In addition to funding provided in Phases I-III, the 2011 reauthorization act also allows agencies to award SBIR Phase I and Phase II award recipients up to $5,000 per year for technical assistance, in addition to the amount of the award, or to provide such assistance through an agency-selected vendor.[17] This funding is intended to provide SBIR recipients with technical assistance services, such as access to a network of scientists and engineers engaged in a wide range of technologies or access to technical and business literature available through online databases. These services are provided to help SBIR awardees make better technical decisions, solve technical problems, minimize technical risks, and develop and commercialize new commercial products and processes.[18]

SBIR Eligibility

A small business' eligibility for the SBIR program is contingent on its location, number of employees, ownership characteristics, and other factors. Eligibility to participate in the SBIR program is limited to for-profit U.S. businesses with a location in the United States. Eligible companies must have 500 or fewer employees, including employees of affiliates. The small business must be:

1) more than 50% directly owned and controlled by one or more citizens or permanent resident aliens of the United States, other small business concerns (each of which is more than 50% directly owned and controlled by individuals who are citizens or permanent resident aliens of the United States), or any combination of these; or

2) more than 50% owned by multiple venture capital operating companies, hedge funds, private equity firms,[19] or any combination of these, with no single such firm owning more than 50% of the small business;[20] or
3) a joint venture in which each entity to the joint venture meets the requirements in paragraphs (1) and (2) above.[21]

Agencies are restricted on how much of their SBIR funds they can make available for awards to small businesses that are more than 50% owned by venture capital operating companies, hedge funds, or private equity firms. The NIH, DOE, and NSF may award no more than 25% of the agency's SBIR funds to such small businesses; all other SBIR agency programs are limited to using 15% of their SBIR funds for such awards.

Small businesses that have received multiple prior SBIR/STTR awards must meet certain benchmark requirements for progress toward commercialization to be eligible for a new Phase I award. For both Phase I and Phase II, the principal investigator's primary employment must be with the small business applicant at the time of award and during the conduct of the proposed project.

Generally, R/R&D work under the STTR must be performed in the United States, though agencies may allow a portion of the work to be performed or obtained outside of the United States under "rare and unique" circumstances.[22]

Current and Historical SBIR Awards Data[23]

In FY2011, the latest year for which the SBA has published data on SBIR awards, agencies made awards for $2.222 billion, including 3,739 Phase I awards totaling $525.4 million and 1,759 Phase II awards totaling $1.696 billion. The success rate[24] was 15% for Phase I SBIR proposers and 49% for Phase II proposers.

While more than two-thirds of SBIR grants made in FY2011 were Phase I awards (68.0%), more than three-fourths of SBIR funding went to Phase II awards (76.4%).[25] Between FY2000 and FY2011, funding for both Phase I and Phase II has generally increased. See *Figure 2*.

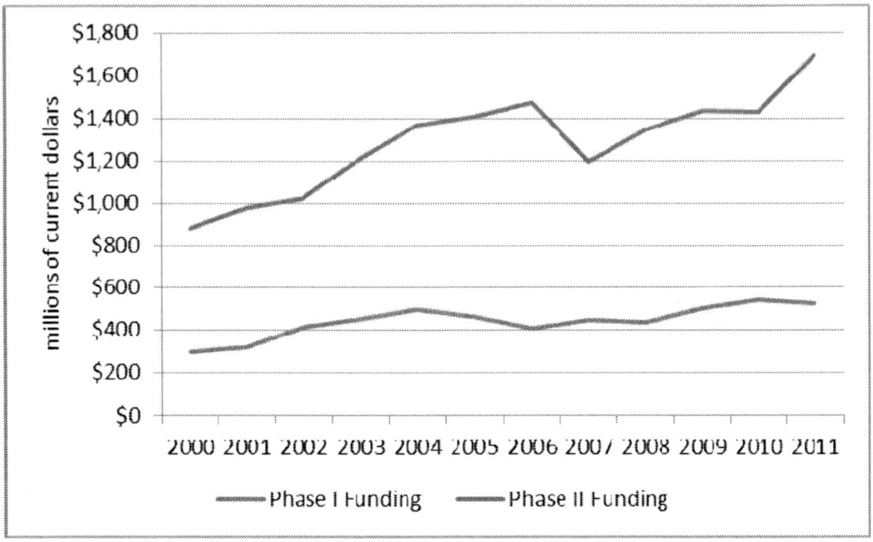

Source: CRS analysis of SBIR.gov annual report data (FY2000-2010) and Small Business Innovation Research (SBIR) and Small Business Technology Transfer (STTR) Program Annual Report FY2009-FY2011 (FY2011).

Figure 2. SBIR Phase I and Phase II Funding, FY2000-FY2011.

Two agencies accounted for more than three-fourths of total SBIR funding in FY2011: DOD ($1,080.8 million, 49%) and HHS ($623.8 million, 28%). The next three highest SBIR funding agencies (NASA, DOE, NSF) together accounted for 20%. The remaining agencies accounted for less than 4% of the total. See *Figure 3*.

The allocation of SBIR funding between Phase I and Phase II awards varies among agencies. Agencies that allocated the largest share of their SBIR funding to Phase I awards in FY2011 were EPA (45%), NSF (36%), and DOC (35%). Agencies that allocated the largest share of their SBIR funding to Phase II awards in FY2011 were DOT (84%), DOD (81%), and DOE (81%). *Figure 4* illustrates each SBIR agency's FY2011 distribution of SBIR funding between phases.

Agency shares of aggregate Phase I and Phase II SBIR funding are as shown in *Figure 5.* The agencies with the highest share of total Phase I funding in FY2011 were DOD (40%), HHS (35%), and NASA (9%). The agencies with the highest share of total Phase II funding in FY2011 were also DOD (51%), HHS (26%), and NASA (8%).

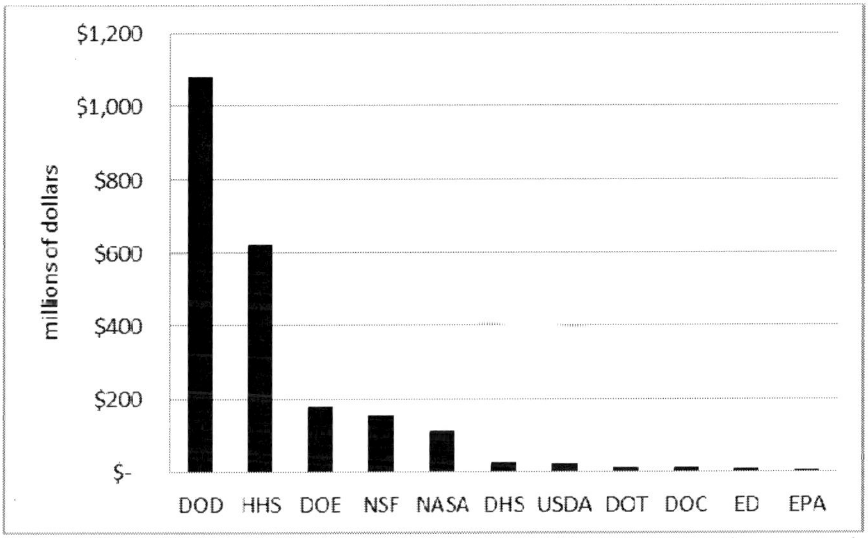

Source: CRS analysis of FY2011 data from Small Business Innovation Research (SBIR) and Small Business Technology Transfer (STTR) Program Annual Report FY2009-FY2011.

Figure 3. SBIR Funding by Agency, FY2011.

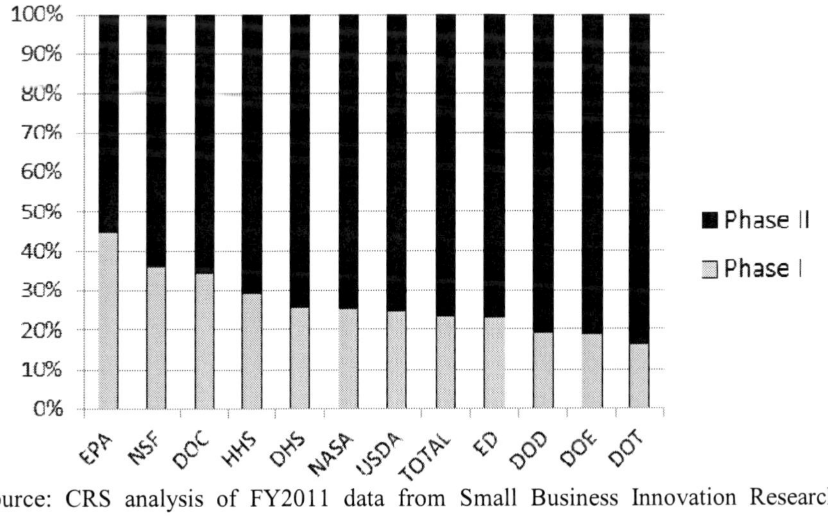

Source: CRS analysis of FY2011 data from Small Business Innovation Research (SBIR) and Small Business Technology Transfer (STTR) Program Annual Report FY2009-FY2011.

Figure 4. Agency Allocation of SBIR Funding Between Phase I and Phase II, FY2011.

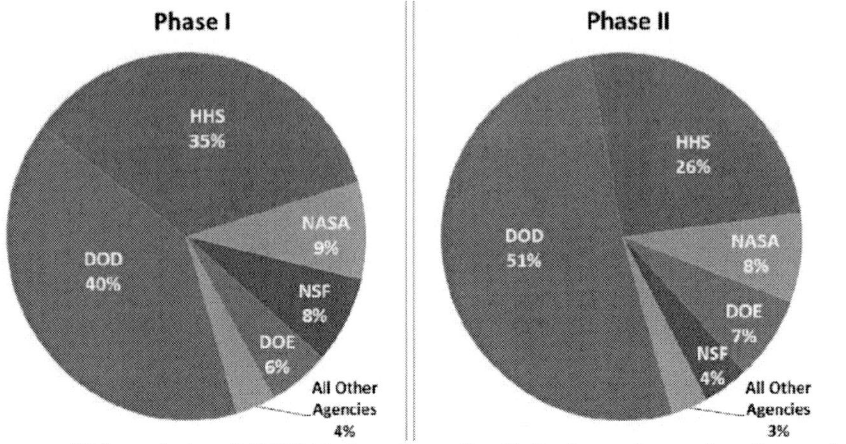

Source: CRS analysis of FY2011 data from Small Business Innovation Research (SBIR) and Small Business Technology Transfer (STTR) Program Annual Report FY2009-FY2011.

Figure 5. Share of Phase I and Phase II SBIR Funding, by Agency, FY2011.

In FY2011, minority or disadvantaged businesses[26] received 238 Phase I awards (about 6.4% of all Phase I SBIR awards) totaling $31.3 million (about 6.0% of total Phase I funding), and 107 Phase II SBIR awards (6.1%) totaling $90.8 million (8.5%). Companies in Historically Underutilized Business Zones (HUBZones)[27] received 87 Phase I awards (about 2.3% of all Phase I awards) totaling $11.5 million (about 2.2% of total Phase I funding) and 45 Phase II awards (2.6%) totaling $29.8 million (2.8%).

Figure 6 shows the aggregate funding level and number of SBIR awards by state for FY2006- 2010 (the latest five-year period for which award data by state are available). Although every state, the District of Columbia, and Puerto Rico received awards during this period, SBIR funding was concentrated among certain states. The four states that received the largest number and amount of SBIR awards during this period—California (5,467 awards totaling $1.826 billion), Massachusetts (3,570 awards totaling $1.235 billion), Virginia (1,786 awards totaling $561.8 million), and Maryland (1,404 awards totaling $492.8 million)—accounted for 38% of the total number of SBIR awards and 43% of the total funding for this period. These four states also received the largest overall amounts of federal R&D funding in FY2010, accounting for a total of 44%.[28]

The top ten states accounted for more than two-thirds of SBIR awards and funding. This concentration mirrors overall federal R&D funding as well. Nine of the top 10 states in SBIR funding are also among the top 10 states in overall federal R&D funding (which account for 66% of total federal R&D funding). In contrast, the ten states with the fewest number of SBIR awards and lowest aggregate award amounts accounted for about 1% of awards and total funding during this period. The ten states with the least federal R&D funding in FY2010 also accounted for about 1% of total federal R&D funding.

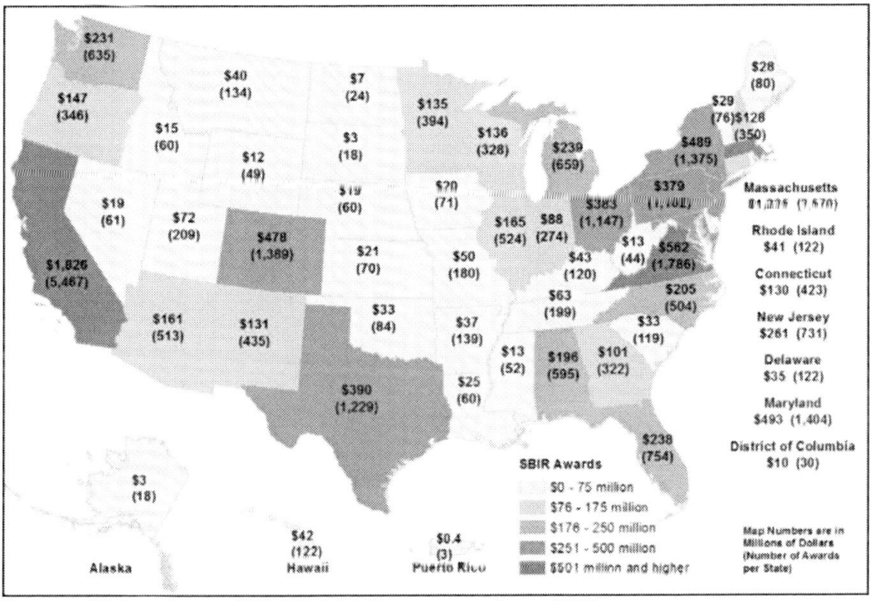

Source: CRS analysis of FY2006-2010 state data from SBIR.gov.

Figure 6. SBIR Aggregate Funding Level and Number of Awards by State, FY2006-2010.

Table 1 provides information on overall agency SBIR obligations for FY2011, as well as the number and aggregate amounts of Phase I and Phase II SBIR awards.

Table 1. Number and Amount of SBIR Awards by Agency, FY2011 (dollars in millions)

Department/Agency	Total Awarded, Phase I and Phase II[a]	Phase I		Phase II	
		Number of Awards	Total Awarded	Number of Awards	Total Awarded[b]
Department of Agriculture	$ 22.4	56	$ 5.5	37	$ 16.9

Table 1. (Continued)

Department/Agency	Total Awarded, Phase I and Phase II[a]	Phase I		Phase II	
		Number of Awards	Total Awarded	Number of Awards	Total Awarded[b]
Department of Commerce	6.1	23	2.1	12	4.0
Department of Defense	1,080.8	1,816	208.1	938	872.7
Department of Education	11.1	25	2.5	12	8.6
Department of Energy	154.0	198	29.3	114	124.7
Dept. of Health and Human Services	623.8	813	183.8	292	440.0
Department of Homeland Security	19.3	44	5.0	18	14.3
Department of Transportation	10.5	15	1.7	12	8.8
Environmental Protection Agency	4.6	27	2.1	11	2.5
Nat'l Aeronautics and Space Admin.	177.6	450	44.8	215	132.8
National Science Foundation	111.5	272	40.4	98	71.1
Total, All Agenciesa	2,221.9	3,739	525.4	1,759	1,696.5

Source: CRS analysis of FY2011 data from Small Business Innovation Research (SBIR) and Small Business Technology Transfer (STTR) Program Annual Report FY2009-FY2011.

[a] Components many not sum to totals due to rounding.

[b] Includes Phase II initial awards, subsequent Phase II awards, and modifications to Phase II awards.

Table 2 provides historical data on the number and amount of Phase I and Phase II SBIR awards from the program's inception through FY2011.

Table 2. Number and Amount of SBIR Awards by Year, FY1983-FY2011

Fiscal Year	Total Dollars		Number of Awards	
	Awarded (in millions)	Phase I	Phase II	Total
FY1983	$ 44.5	686	74	760
FY1984	108.4	999	338	1,337
FY1985	199.1	1,397	407	1,804
FY1986	297.9	1,945	564	2,509
FY1987	350.5	2,189	768	2,957
FY1988	389.1	2,013	711	2,724
FY1989	431.9	2,137	749	2,886

Fiscal Year	Total Dollars		Number of Awards	
	Awarded (in millions)	Phase I	Phase II	Total
FY1990	460.7	2,346	837	3,183
FY1991	483.1	2,553	788	3,341
FY1992	508.4	2,559	916	3,475
FY1993	698.0	2,898	1,141	4,039
FY1994	717.6	3,102	928	4,030
FY1995	981.7	3,085	1,263	4,348
FY1996	916.3	2,841	1,191	4,032
FY1997	1,066.7	3,371	1,404	4,775
FY1998	1,100.0	3,022	1,320	4,342
FY1999	1,096.5	3,334	1,256	4,590
FY2000	1,190.2	3,172	1,335	4,507
FY2001	1,294.4	3,215	1,533	4,748
FY2002	1,434.8	4,243	1,577	5,820
FY2003	1,670.1	4,465	1,759	6,224
FY2004	1,867.4	4,638	2,013	6,651
FY2005	2,029.8	4,300	1,871	6,171
FY2006	2,113.9	3,836	2,026	5,862
FY2007	1,777.6	3,909	1,615	5,356
FY2008	1,785.7	3,832	1,851	5,683
FY2009	1,937.7	4,008	1,801	5,809
FY2010	1,970.5	4,146	1,845	5,991
FY2011	2,221.9	3,739	1,759	5,498

Source: U.S. Small Business Administration, The Small Business Economy 2010, A Report to the President, http://www.sba.gov/sites/default/files/sb_econ2010.pdf; SBIR.gov website; and Small Business Innovation Research (SBIR) and Small Business Technology Transfer (STTR) Program Annual Report FY2009-FY2011.

SMALL BUSINESS TECHNOLOGY TRANSFER

STTR Overview

The Small Business Technology Transfer (STTR) program was created by the Small Business Research and Development Enhancement Act of 1992 (P.L. 102-564) and has been reauthorized several times, most recently by the SBIR/STTR Reauthorization Act of 2011 (P.L. 112-81) which reauthorized the program through September 30, 2017. Modeled largely after the SBIR program, the STTR program seeks to facilitate the commercialization of university and federal R&D by small companies. Under the program, each federal agency with extramural R&D budgets of $1 billion or more is required to allocate a portion of its R&D funding to conduct a multi-phase R&D grant program for small businesses. The STTR program provides funding for research proposals that are developed and executed cooperatively between a small firm and a scientist in an eligible research institution[29] and that are aligned with the mission requirements of the federal funding agency.

Currently, five agencies participate in the STTR program: DOD, DOE, HHS, NASA, and NSF. Under the 2011 reauthorization act, the minimum percentage of funds to be set aside for the program is to increase from 0.30% in FY2011 to 0.35% in FY2012 and FY2013; to 0.40% in FY2014 and FY2015; and to 0.45% in FY2016 and beyond. In FY2011, total STTR award funding among all STTR-participating federal agencies was $238.1 million, accounting for 0.31% of the agencies' aggregate extramural R&D funding.

The SBA emphasizes three principal differences between the STTR and SBIR programs:

- Under STTR, the small business and its partnering research institution must establish an intellectual property agreement detailing the allocation of intellectual property rights and rights to carry out follow-on research, development or commercialization activities.
- Under STTR, the small business partner must perform at least 40% of the R&D and the research institution partner must perform at least 30% of the R&D.
- The STTR program does not require the principal investigator to be primarily employed by the small business, a requirement of the SBIR program.[30]

As with the SBIR program, each participating agency operates its own STTR program under the provisions of the law and regulations, as well as with the policy directive issued by the U.S. Small Business Administration (SBA) in its *Small Business Technology Transfer Program Policy Directive* (referred to hereinafter as the *STTR Program Policy Directive*).[31] According to some analysts, this approach allows for general consistency across STTR programs, while allowing each agency a substantial degree of control and flexibility in the execution of its program in alignment with its overall mission and priorities.[32] (See "Improving Technology Commercialization, Trade-Offs Among Program Objectives" for related discussion.)

STTR Phases

Like the SBIR program, the STTR program has three phases. The purposes and parameters of each phase are discussed below.

Phase I

In Phase I, an agency solicits contract proposals or grant applications to conduct feasibility-related experimental or theoretical research or research and development (R/R&D) related to agency requirements. The scope of the topic(s) in the solicitation may be broad or narrow, depending on the needs of the agency. Phase I grants are intended to determine "the scientific and technical merit and feasibility of the proposed effort and the quality of performance of the [small business] with a relatively small agency investment before consideration of further Federal support in Phase II."[33] Generally, STTR Phase I awards are limited to $150,000 (the Phase I award guideline), though law provides agencies with the authority to issue awards that exceed this guideline by as much as 50%. In addition, agencies may request a waiver from the SBA to exceed the award guideline by more than 50% for a specific topic. In general, the period of performance for Phase I awards is not to exceed one year, though agencies may allow for a longer performance period for a particular project.

Phase II

Phase II grants are intended to further R/R&D efforts initiated in Phase I that meet particular program needs and that exhibit potential for commercial application. In general, only Phase I grant recipients are eligible Phase I and the scientific and technical merit and commercial potential of the project proposed in Phase II. The *STTR Program Policy Directive* generally limits STTR Phase II awards to $1 million (the Phase II award guideline), though the directive provides agencies with the authority to issue awards that exceed this guideline by as much as 50% (for an amount up to $1.5 million). As with Phase I grants, agencies may request a waiver from the SBA to exceed the Phase II award guideline by more than 50% for a specific topic. In general, the period of performance for Phase II awards is not to exceed two years, though agencies may allow for a longer performance period for a particular project. Agencies may make a sequential Phase II award to continue the work of an initial Phase II award. This sequential Phase II award is also subject to the $1 million Phase II guideline and agencies' authority to exceed the guideline by up to 50%. Thus, agencies may award up to $3 million in Phase II awards for a particular project to a single recipient at the agency's discretion, and potentially more if the agency requests and receives a waiver from the SBA. For sequential Phase II awards, some agencies require third-party matching of the STTR funds.

Phase III

Phase III of the STTR program is focused on the commercialization of the results achieved through Phase I and Phase II STTR funding. The STTR program **does not** provide funding in Phase III. Phase III funding is expected, generally, to be generated in the private sector. However, some agencies may use non-STTR funds for Phase III funding to support additional R&D or contracts for products, processes, or services intended for use by the federal government. In addition, the 2011 reauthorization act directs agencies and prime contractors "to the greatest extent practicable," to facilitate the commercialization of SBIR and STTR through the use of Phase III awards, including sole source awards.[35]

Technical Assistance

The 2011 reauthorization act also allows agencies to award STTR Phase I and Phase II award recipients up to $5,000 per year for technical assistance, in addition to the amount of the award, or to provide such assistance through a vendor.[36] This funding is intended to provide STTR recipients with technical assistance services, such as access to a network of scientists and engineers engaged in a wide range of technologies or access to technical and business literature available through online databases. These services are provided to help STTR awardees make better technical decisions, solve technical problems, minimize technical risks, and develop and commercialize new commercial products and processes.[37]

STTR Eligibility

A small business' eligibility for the STTR program is contingent on its location, number of employees, ownership characteristics, and other factors. The partnering research institution must meet eligibility qualifications as well. Eligibility to participate in the STTR program is limited to for-profit U.S.

businesses with a location in the United States. Eligible companies must have 500 or fewer employees, including employees of affiliates.

The small business must be:

1) more than 50% directly owned and controlled by one or more citizens or permanent resident aliens of the United States, other small business concerns (each of which is more than 50% directly owned and controlled by individuals who are citizens or permanent resident aliens of the United States), or any combination of these; or
2) a joint venture in which each entity to the joint venture meets the requirements in paragraph (1) above.[38]

Unlike the SBIR program, the STTR does not have authority to make awards to small businesses that are more than 50% owned by multiple venture capital operating companies, hedge funds, private equity firms, or any combination of these. However, as with SBIR, the STTR program may make awards to companies that are majority-venture capital backed if the VC firm is itself more than 50% directly owned and controlled by one or more individuals who are citizens or permanent resident aliens of the United States. In such a case, that VC is allowed to have majority ownership and control of the awardee; however, the VC and the awardee, and all other affiliates, must have a total of 500 employees or less.[39]

In addition, small businesses that have received multiple prior SBIR/STTR awards must meet certain benchmark requirements for progress toward commercialization to be eligible for a new Phase I award. For both Phase I and Phase II, the principal investigator's primary employment must be with either the small business or the partnering research institution at the time of award and during the conduct of the proposed project. Generally, R/R&D work under the STTR must be performed in the United States, though agencies may allow a portion of the work to be performed or obtained outside of the United States under "rare and unique" circumstances.[40]

The partnering research institution must be located in the United States, and be either a nonprofit college or university, a domestic nonprofit research organization,[41] or a federally funded research and development center (FFRDC).[42]

For both Phase I and Phase II, not less than 40% of the R/R&D work must be performed by the small business, and not less than 30% of the R/R&D work must be performed by the single, partnering research institution. Agencies can choose whether to determine these percentages using either contract dollars or labor hours, but must explain this in the solicitation.

Current and Historical STTR Awards Data[43]

In FY2011, the most recent year for which the SBA has published data on STTR awards, agencies made awards for $251.2 million, including 482 Phase I STTR awards totaling $59.6 million and 238 Phase II STTR awards totaling $191.6 million. The success rate for Phase I STTR proposers was 22% and for Phase II proposers was 44%.

While 67% of STTR grants made in FY2011 were for Phase I awards, more than 76% of STTR funding went to Phase II awards. In FY2004, the STTR set-aside doubled from 0.15% to 0.30%. In the first year (FY2004), aggregate funding for Phase I and aggregate funding for Phase II approximately doubled. However, from FY2004 to FY2011, Phase I aggregate funding fell by about 25% while Phase II aggregate funding increased by about 74%. The proportional change in funding between the phases may reflect an increased focus on commercialization by the STTR agencies. See *Figure 7* for Phase I and Phase II STTR funding for FY2000-2011.

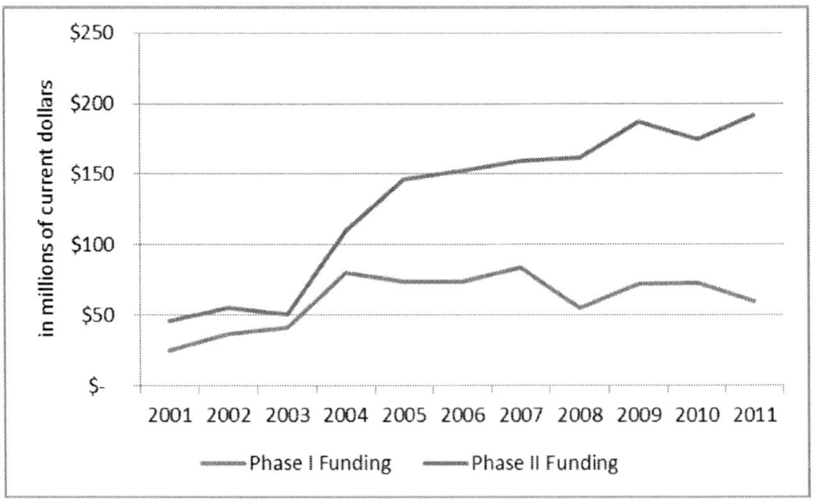

Source: CRS analysis of SBIR.gov annual report data (FY2000-2010) and Small Business Innovation Research (SBIR) and Small Business Technology Transfer (STTR) Program Annual Report FY2009-FY2011 (FY2011).

Figure 7. STTR Phase I and Phase II Funding, FY2000-FY2011.

Like SBIR funding, STTR funding is highly concentrated. Two agencies—DOD ($120.9 million, 48%) and HHS ($77.5 million, 31%)—accounted for nearly four-fifths of STTR funding in FY2011. NASA accounted for 8%, DOE for 7%, and NSF for 5%. See *Figure 8*.

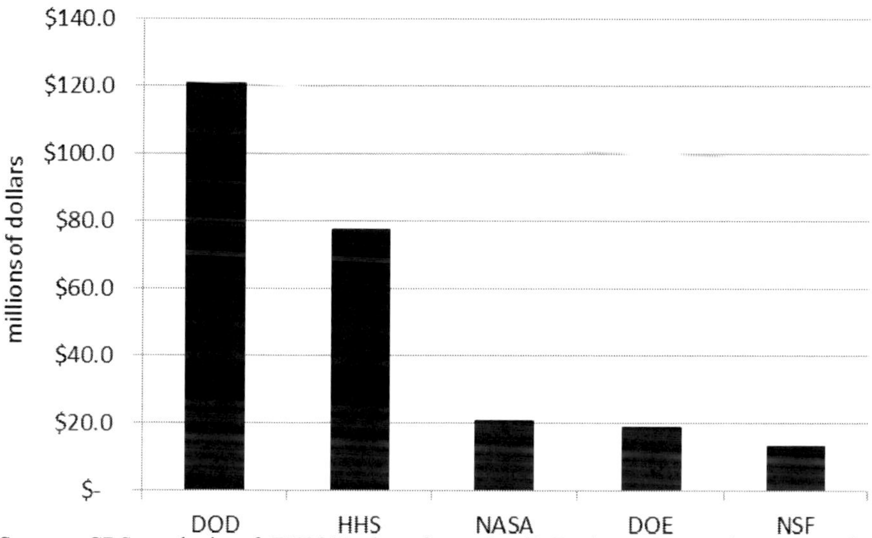

Source: CRS analysis of FY2011 data from Small Business Innovation Research (SBIR) and Small Business Technology Transfer (STTR) Program Annual Report FY2009-FY2011.

Figure 8. STTR Funding by Agency, FY2011.

The allocation of STTR funding to Phase I and Phase II awards varies among agencies, but the differences are smaller than for SBIR funding. Among STTR agencies, HHS allocated the largest share (29%) of its STTR funding to Phase I awards in FY2011; NSF allocated the largest share (95%) to Phase II awards. See *Figure 9*.

The agencies with the highest share of total Phase I funding in FY2011 were DOD (50%) and HHS (37%). The agencies with the highest share of total Phase II funding in FY2011 were also DOD (48%) and HHS (29%). See *Figure 10*.

Minority or disadvantaged businesses[44] received 31 Phase I STTR awards (6% of all Phase I STTR awards) totaling $3.1 million (5% of total Phase I STTR funding) in FY2011, and 14 Phase II STTR awards (6%) totaling $7.9 million (4%).

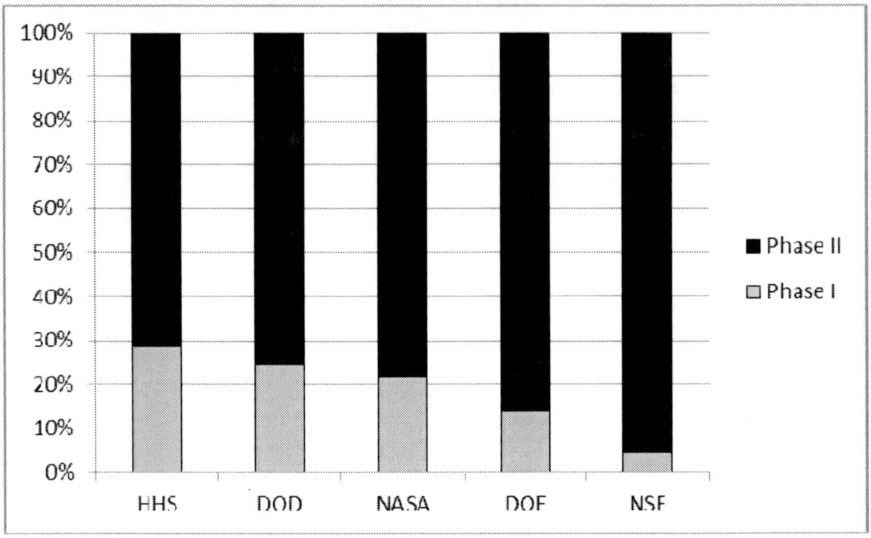

Source: CRS analysis of FY2011 data from Small Business Innovation Research (SBIR) and Small Business Technology Transfer (STTR) Program Annual Report FY2009-FY2011.

Figure 9. Agency Allocation of STTR Funding Between Phase I and Phase II, FY2011.

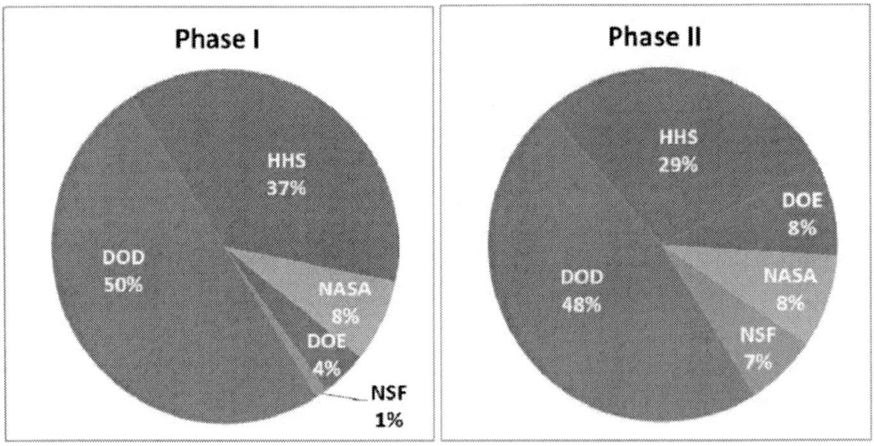

Source: CRS analysis of FY2011 data from Small Business Innovation Research (SBIR) and Small Business Technology Transfer (STTR) Program Annual Report FY2009-FY2011.

Figure 10. Share of Phase I and Phase II STTR Funding, by Agency, FY2011.

Figure 11 shows the aggregate funding level and number of STTR awards by state for FY2006- FY2010 (the latest five-year period for which award data by state are available). STTR funding was concentrated in certain states. The three states that received the largest number and amount of STTR awards during this period—California (651 awards totaling $195.8 million), Massachusetts (474 awards totaling $139.9 million), and Virginia (267 awards totaling $77.2 million)— accounted for 33% of the total number of SBIR awards and 32% of the total funding for this period. The top ten states accounted for more than 60% of awards and funding. In contrast, the ten states with the fewest awards and lowest aggregate award amounts accounted for about 1% of awards and total funding during this period.

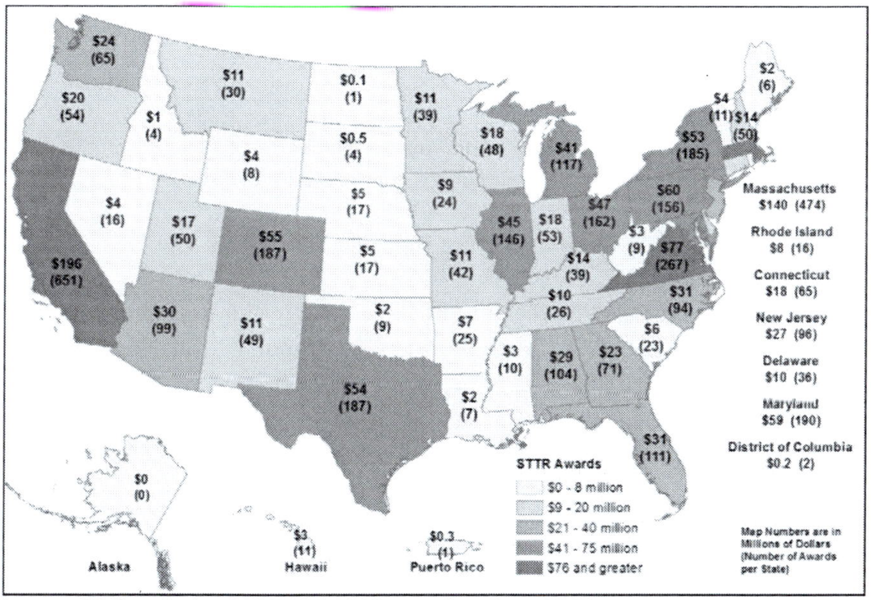

Source: CRS analysis of FY2006-2010 state data from SBIR.gov.

Figure 11. STTR Aggregate Funding Level and Number of Awards by State, FY2006-2010.

Table 3 provides information on overall agency STTR obligations for FY2011, as well as the number and aggregate amounts of Phase I and Phase II awards.

Table 3. Number and Amount of STTR Awards by Agency, FY2011 (in millions of dollars)

		Phase I		*Phase II*	
Department/Agency	*Total Amount Awarded, Phase I and Phase IIa*	*Number of Awards*	*Total Amount Awarded*	*Number of Awards*	*Total Amount Awarded*
Department of Defense	$ 121.0	309	$ 29.8	127	$ 91.2
Department of Energy	18.8	26	2.6	22	16.2
Dept. of Health and Human Services	77.5	98	22.1	44	55.4
Nat'l Aeronautics and Space Admin.	20.7	45	4.5	27	16.2
National Science Foundation	13.2	4	0.6	18	12.6
Total, All Agencies[a]	*251.2*	*482*	*59.6*	*238*	*191.6*

Source: CRS analysis of FY2011 data from Small Business Innovation Research (SBIR) and Small Business Technology Transfer (STTR) Program Annual Report FY2009-FY2011.

[a] Components many not sum to totals due to rounding.

Table 4 provides historical information on the number of Phase I and Phase II STTR awards and total annual STTR funding from the program's inception through FY2011.

Table 4. Number and Amount of STTR Awards by Year, FY1994-FY2011

	Dollars Awarded (in millions)			Number of Awards		
Fiscal Year	Phase I	Phase II	Total	Phase I	Phase II	Total
FY1994	$18.9	$—	$ 18.9	198	—	198
FY1995	23	10.7	33.7	238	22	260
FY1996	22.7	41.8	64.5	238	88	326
FY1997	24.2	44.9	69.1	260	89	349
FY1998	19.7	45.1	64.8	208	109	317
FY1999	24.3	40.6	64.9	251	78	329
FY2000	23.9	45.9	69.8	233	95	328
FY2001	24.2	53.2	77.4	224	113	337
FY2002	36.4	55.4	91.8	356	114	470
FY2003	41.1	50.7	91.8	397	111	508
FY2004	79.7	110.3	190.0	674	195	869
FY2005	73.9	146.4	220.3	611	221	832

	Dollars Awarded (in millions)			Number of Awards		
Fiscal Year	Phase I	Phase II	Total	Phase I	Phase II	Total
FY2006	74.0	152.3	226.3	644	234	878
FY2007	83.5	159.4	242.9	634	213	847
FY2008	61.2	178.4	239.6	483	251	734
FY2009	72.1	186.9	259.0	592	251	843
FY2010	73.1	174.3	247.4	625	256	881
FY2011	59.6	191.6	251.2	482	238	720

Source: U.S. Small Business Administration, The Small Business Economy 2010, A Report to the President, http://www.sba.gov/sites/default/files/sb_econ2010.pdf; SBIR.gov website; and Small Business Innovation Research (SBIR) and Small Business Technology Transfer (STTR) Program Annual Report FY2009-FY2011.

SBIR/STTR Reauthorization Act of 2011 Provisions

The SBIR/STTR Reauthorization Act of 2011 (enacted as Division E of the National Defense Authorization Act for Fiscal Year 2012, P.L. 112-81) authorizes the SBIR and STTR programs through September 30, 2017. The act also changes certain aspects of the programs. This section provides an overview of these changes.

Participation of Small Businesses That Are Majority-Owned by Venture Capital Companies

Perhaps the most widely debated issue of the reauthorization was whether to permit small companies that are majority-owned by venture capital operating companies, hedge funds, or private equity firms to receive grants under the SBIR and STTR programs. In what might be considered a compromise position, the act permits NIH, DOE, and NSF to award not more than 25% of SBIR funds to small businesses "that are owned in majority part by multiple venture capital operating companies, hedge funds, or private equity firms through competitive, merit-based procedures that are open to all eligible small business concerns."[46] Other federal agencies may not award more than 15% of SBIR funds to such firms. The act directs the GAO to conduct triennial studies on venture capital operating company, hedge fund, and private equity firm involvement in the program. The first report is due in

December 2014. For further discussion of this issue, see "Eligibility of Venture Capital-Backed Small Businesses."

Agency Set-Aside Percentages and Award Limitations

The act increases the percentages of extramural R&D funding that agencies must set aside for the SBIR and STTR programs, introducing the changes over multiple years. See *Table 5.*

Table 5. SBIR and STTR Set-Aside Percentages by Year

	SBIR Set-Aside Percentage	STTR Set-Aside Percentage
FY2011	2.5%	0.30%
FY2012	2.6%	0.35%
FY2013	2.7%	0.35%
FY2014	2.8%	0.40%
FY2015	2.9%	0.40%
FY2016	3.0%	0.45%
FY2017 and later fiscal years	3.2%	0.45%

Source: P.L. 112-81.

Additionally, the law increases the award guidelines on Phase I SBIR/STTR awards from $100,000 to $150,000 and on Phase II SBIR/STTR awards from $750,000 to $1,000,000. Agencies cannot exceed these guidelines by more than 50% without a waiver from the SBA. The act also provides express authority to agencies to make a sequential Phase II award to continue the work of an initial Phase II award. Sequential Phase II awards are also subject to the $1 million guideline/$1.5 million limit. Upon agency request, the SBA Administrator may grant a waiver allowing an agency to exceed the limits with respect to a specific topic for a fiscal year if the limitations will interfere with the ability of the agency to fulfill its research mission through the SBIR program or the STTR program. The agency must agree to minimize the number of awards that exceed the award guidelines.

GAO Audit of Agencies' Extramural R&D Budget Calculations

The act directs the GAO to audit and report on agency calculation of their extramural R&D budgets. GAO has reported that agencies have been inconsistent and late in reporting to the SBA their explanations of how they calculate their extramural R&D budgets, which are the basis used to calculate the minimum SBIR and STTR set-aside amounts.[47] For further discussion of this issue, see "Calculation of Extramural Research Funding and Set-Aside."

Company Flexibility in Pursuit of Phase II Grants

The act gives small businesses more flexibility in applying for Phase II awards. Recipients of a Phase I award from one federal agency may now apply for a Phase II award from another agency to pursue the original work (e.g., a company that received a Phase I award from the Department of Energy may apply for a Phase II award from the Department of Defense to build on its Phase I work). In addition, a small business may switch between the SBIR and STTR programs for Phase I and Phase II awards (e.g., a small business that wins an SBIR Phase I award may now compete for a Phase II STTR award). The act requires agency heads to verify that any activity to be performed with respect to a project with a Phase I or Phase II SBIR or STTR award has not been funded under the SBIR program or STTR program of another Federal agency to prevent the duplication of funded work.[48]

In addition, the act establishes a pilot program that allows the Department of Defense, Department of Education, and National Institutes of Health to award Phase II grants to small businesses that did not first receive a Phase I grant.

Commercialization Focus

The 2011 reauthorization act includes a number of provisions seeking to increase the programs' effectiveness in technology commercialization. The act requires each SBIR/STTR agency to establish a system to measure the success of small businesses that received Phase I awards in securing Phase II awards. Agencies are also required to establish minimum performance standards for small businesses in advancing from a Phase I award to a Phase II award, and to evaluate each recipient's compliance with the standard. Small firms that fail to meet this standard are barred from competing for additional Phase I SBIR or STTR awards from that agency for a one year. Similarly, the act requires each agency to: establish systems to measure the success of SBIR and STTR awardees in securing Phase III SBIR or STTR awards, establish a minimum performance standard in this regard, evaluate each recipient's compliance with the standard, and bar firms that fail to meet this standard from competing for additional Phase I (and in some cases, Phase II) SBIR or STTR awards for one year. Agencies are required to report their tracking systems and minimum performance standards to the SBA Administrator for approval.

The act also authorized agencies to establish commercialization readiness pilot programs. This authority allows each agency to use up to 10% of its SBIR and STTR funds to make awards of up to three times the dollar amounts established for Phase II awards. These awards may be used to support technology development, testing, evaluation, and commercialization assistance for SBIR and STTR Phase II technologies, or to support the progress of R/R&D and commercialization conducted under the SBIR or STTR programs to Phase III. To establish a commercialization readiness pilot program, agencies must first make a written application to the SBA Administrator for approval describing

> a compelling reason that additional investment in SBIR or STTR technologies is necessary, including unusually high regulatory, systems integration, or other costs relating to development or manufacturing of identifiable, highly promising small business technologies or a class of such technologies expected to substantially advance the mission of the agency.[49]

In making such awards, agency heads are directed to consider whether the technology to be supported by the award is likely to be manufactured in the United States.

The act encourages agencies to award SBIR and STTR grants to small businesses that work with federal laboratories or that are involved in cooperative research and development agreements (also known as CRADAs).

In addition, the act allows agencies to contract with a vendor to provide SBIR/STTR awardees with technical assistance services. Such services could include access to a network of scientists and engineers engaged in a wide range of technologies or access to technical and business literature available through online databases. Funding of these services is intended to help the small businesses make better technical decisions, solve technical problems which arise during the conduct of their SBIR/STTR projects, minimize technical risks associated with such projects, and develop and commercialize new commercial products and processes resulting from such projects. Alternatively, an agency may authorize SBIR/STTR awardees to purchase such services up to $5,000 per year, in addition to the amount of the recipient's award.

The act also establishes a "Phase 0 Proof of Concept Partnership Pilot Program" at NIH to accelerate the creation of small businesses and the commercialization of research innovations from universities or other research institutions that participate in the NIH STTR program. Under this pilot, NIH may make awards of up to $1 million per year for up to three years. These funds may be used to support technical validations, for market research, to clarify intellectual property rights position and strategy, or to investigate commercial or business opportunities. These funds may not be used for basic research activities or for the acquisition of research equipment or supplies unrelated to commercialization activities.

Another commercialization-focused provision of the act provides a special acquisition preference to SBIR and STTR award recipients. The act directs agencies and prime contractors, to the greatest extent practicable, to issue Phase III awards relating to technology, including sole source awards, to the SBIR and STTR award recipients that developed the technology.

Annual Reporting of Agencies' Advanced Manufacturing Activities

The act requires each agency that makes total SBIR and STTR awards in excess of $50 million to report annually to the SBA Administrator on efforts to improve U.S. manufacturing activities and to make recommendations for further improvements. The SBA is required to incorporate the agency reports into its mandatory annual report to Congress.

Data Rights Protection

The act includes a provision to protect the rights of small businesses to data generated in the performance of an SBIR award for a period of not less than four years. In addition, the act directs GAO to report to Congress on the implementation and effectiveness of data rights protection. Specifically, the act directs GAO to assess whether federal agencies comply with data rights protections for SBIR awardees and their technologies;[50] whether the laws and policy directives intended to clarify the scope of data rights are sufficient to protect SBIR awardees; and whether there is an effective grievance tracking process for SBIR awardees who have grievances against a federal agency regarding data rights and a process for resolving those grievances. The act required a report within 18 months of its enactment (approximately June/July 2013). GAO issued a letter in November 2013 stating that it was awaiting SBA's revision of the policy directive as it "has a bearing on the issue of whether laws and policy directives are sufficient to protect SBIR awardees."[51] SBA subsequently published its updated policy directive on February 24, 2014. GAO has not published a report on this matter as of August 2014.

Provisions to Reduce Waste, Fraud, and Abuse

Congress has expressed continuing concerns about waste, fraud, and abuse in the SBIR and STTR programs. The 2011 reauthorization act includes a number of provisions to identify and eliminate waste, fraud, and abuse. To this end, the act:

- requires the SBA administrator to amend the SBIR Policy Directive and the STTR Policy Directive to include measures to prevent fraud, waste, and abuse;
- directs that the amendments to the policy directives include definitions or descriptions of fraud, waste, and abuse; guidelines for the monitoring and oversight of applicants to, and recipients of, awards; and a requirement that each SBIR/STTR agency provide information on the method established by each agency inspector general to report fraud, waste, and abuse on its website and in any SBIR/STTR solicitation;
- requires SBIR and STTR applicants and award recipients to certify its compliance with the laws relating to the programs and the conduct guidelines established under the policy directives;
- directs inspectors general in SBIR and STTR agencies to establish fraud detection indicators; review regulations and operating procedures; coordinate information sharing between agencies, to the extent otherwise permitted under federal law; and improve the education and training of and outreach to program administrators, applicants, and recipients; and
- requires the GAO to publish an initial report within one year from the date of enactment and every four years thereafter.

For further discussion of this issue, see "Concerns About Duplicative Awards and Other Types of Waste, Fraud, and Abuse."

Management and Administrative Improvements

The act requires the Director of the Office of Science and Technology Policy to establish an Interagency SBIR/STTR Policy Committee that includes representatives of the SBA and all agencies with an SBIR or STTR program. The law directs the committee to develop policy recommendations on ways to improve program effectiveness and efficiency, including issues related to development of the SBIR and STTR awards databases; agency flexibility in establishing Phase I and II award sizes; best practices in technology commercialization and mechanisms for addressing company funding gaps after Phase II but prior to commercialization; a framework for a systematic assessment of SBIR and STTR programs, including tracking awards and outcomes; and outreach and technical assistance activities that increase the participation of small businesses underrepresented in the SBIR and STTR programs. Following initial one-year and 18-month reports, the committee is to report to selected committees of Congress every two years.

The act also establishes a pilot program that allows agencies to use no more than 3% of SBIR program funds for administrative activities, oversight, and contract processing. Among the authorized uses of these funds are: to support the administration of the SBIR and STTR programs; to support outreach and technical assistance relating to the SBIR and STTR programs, including technical assistance site visits, personnel interviews, and national conferences; to increase outreach activities to increase the participation of women-owned and socially and economically disadvantaged small business concerns; to support the implementation of commercialization and outreach initiatives of P.L. 112-81; to increase participation of states that have traditionally received low levels of SBIR awards; to support activities related to congressional oversight, including the prevention of waste, fraud, and abuse; to carry out the laws authorizing participation by majority venture capital-owned small businesses; to pay for contract processing costs relating to the SBIR and STTR programs; and to pay for additional personnel and assistance with application reviews.

The act further required the SBA to publish revised SBIR and STTR policy directives incorporating the act's changes in the programs mandated by the act within 180 days of the passage of the legislation. SBA published revised policy directives for comment in the *Federal Register* on August 6, 2012; the comment period closed on October 5th; and final action occurred in December 2013. The policy directives were updated on February 24, 2014. The final rule for venture capital participation was finalized and published in the *Federal Register* on December 27, 2012.[52]

ISSUES FOR CONSIDERATION

As it has since establishing the SBIR and STTR programs, Congress seeks to better understand and address challenges to the programs' effectiveness. The following section provides an overview of selected ongoing issues that Congress may opt to consider.

Eligibility of Venture Capital-Backed Small Businesses

Much of the debate over the reauthorization of the SBIR and STTR programs in 2011 revolved around a regulation that required at least 51% ownership by an individual or individuals. Some experts argued that participation by small firms that are majority-owned by venture capital companies, hedge funds, and private equity firms should be permitted. Proponents of this change maintained that, particularly in the biotechnology sector, the most innovative companies were not able to use the SBIR program because they did not meet these ownership criteria. Opponents of altering the eligibility requirements argued that the program is designed to provide financial assistance where venture capital is not available. They asserted that the program's objective is to bring new concepts to the point where private sector investment is feasible. While the new law permits limited participation by majority venture capital owned companies, it remains to be seen how this will affect the outcomes of the two programs.

Agency Compliance with Mandatory Minimum Expenditure Levels

A continuing issue for the SBIR and STTR programs is agency compliance with expending the statutory minimum percentage of extramural research funding annually. In a September 2013 report, GAO found that 8 of the 11 agencies participating in the SBIR program and 4 of the 5 agencies participating in the STTR program failed to consistently comply with spending requirements for FY2006-FY2011.[53] In June 2014, GAO reported that three agencies failed to comply with the SBIR requirement and three failed to comply with the STTR requirement in FY2012. GAO reported that program managers at two of the non-compliant agencies asserted that their agencies would be in compliance if the agencies spent the total amount reserved or budgeted for their programs, regardless of what year the funding was spent. GAO asserted that the law requires agencies to "expend" a certain amount of funding each year and attributes the agencies' misinterpretation, in part, to the SBA's SBIR and STTR policy directives which "inaccurately state that the authorizing legislation requires agencies to 'reserve' the minimum amount each year."[54]

Among the factors affecting agencies' failure to comply with meeting the mandatory minimum expenditure levels are challenges in calculating the amount to be set aside; the enactment of appropriations after the start of the fiscal year; and differing agency interpretations of the statutory requirement for "expended."

Calculation of Extramural Research Funding and Set-Aside

The SBIR and STTR set-asides are based on an agency's extramural budget for research or research and development.[55] The calculation of the amount of this budget can be complex for some agencies. For example, several agencies support extramural R/R&D funding through multiple subunits.[56] In addition, agency extramural R/R&D funding can come from more than one appropriations account, and such accounts can include activities and programs that are not extramural R/R&D.[57] Accordingly, each agency must determine its extramural R/R&D budgets using a methodology that identifies extramural R/R&D funding as well as what is to be excluded from this amount.[58]

Given the complexity of this challenge, Congress required each agency to report its methodology to SBA annually within four months of enactment of its appropriation.[59] The *SBIR Program Policy Directive* requires that this report also include an itemization and explanation of excluded items.[60] However, GAO also found that at least six agencies did not itemize and/or explain exclusions from their calculations. In addition, according to GAO, agencies generally have submitted these reports to SBA too late for the SBA to provide timely feedback to the agencies after reviewing their methodologies and exclusions. GAO recommended that agencies submit their methodology reports in accordance with the four months provided after enactment of appropriations as specified in law.[61]

Another factor affecting the calculation of SBIR funding is that, in practice, agencies generally calculate their SBIR set-asides based on their extramural R/R&D *budgets* and not on their extramural R/R&D *obligations* as required by statute.[62] An agency's extramural R/R&D budget reflects its spending plans for a fiscal year, whereas an agency's extramural R/R&D obligations reflect the amount of funds an agency obligates[63] to spending in a fiscal year; a final obligation figure for extramural R/R&D may not be calculable until the end (or very close to the end) of a fiscal year. Thus, an agency's extramural R/R&D obligations (and the minimum SBIR set-aside amount) may be higher or lower than the level the agency anticipated in its extramural R/R&D budget.

Enactment of Appropriations after Start of Fiscal Year

Enactment of appropriations after the start of a fiscal year may also affect the ability of agencies to expend SBIR/STTR funds in that fiscal year. For example, if an agency plans its expenditures around a level specified in a continuing resolution but then receives a higher level of funding in its final appropriations act(s), then expenditure of the additional amount set aside for SBIR/STTR in that fiscal year may be difficult.

A related factor that may delay calculation of the amount to be set aside for SBIR/STTR is the time required for an agency to determine the amount of its extramural research funding. Appropriations acts often provide funding to accounts with multiple purposes, including extramural R&D, intramural R&D, and non-R&D activities. In such cases, agencies must make allocation decisions for these funds (subject to limitations and guidance provided in the appropriations act(s) and related report language) before the extramural research budget can be calculated.

Similarly, agency officials have asserted that agencies had already planned their programs and made awards in FY2012 prior to enactment of the SBIR/STTR Reauthorization Act of 2011 which contained provisions increasing the set aside percentage and which was enacted a quarter into FY2012.[64]

Agency Views of Requirement to "Expend" Funds

Some program managers at agencies that fell short of the statutorily required expenditures in FY2012 told GAO that they believed their agency was in compliance if their agency spent the total amount reserved or budgeted for the program regardless of what year the funding is spent. GAO, however, responded that the statute requires each agency to "expend" the funds in the year it is set aside. GAO recommended that SBA revise its SBIR and STTR policy directives to accurately reflect the statutory language regarding program spending requirements.[65]

Congress might consider statutory changes that alter or clarify how agencies are to determine the amount to be set aside each year for SBIR and STTR, and whether those amounts must be spent in the same fiscal year; obligated, in whole or in part, for expenditure over multiple fiscal years; or expended without restriction to any given period.

Improving Technology Commercialization, Trade-Offs Among Program Objectives

A statutory goal of the SBIR and STTR programs is to foster the development and commercialization of new technologies. Success in achieving this goal can take different forms, for example an innovation that addresses an agency need (e.g., an improved material for a NASA spacecraft), a commercial opportunity, or both. Such innovations can promote economic growth, job creation, and national competitiveness, or address other societal needs and challenges such as national defense, public health, and environmental protection. The 2011 reauthorization act includes a number of provisions focused on improving commercialization. For example, the act authorizes agencies to provide assistance to SBIR/STTR awardees to overcome technical barriers and to allow agencies to establish commercialization readiness pilot programs.

Some analysts have cautioned against placing too much emphasis on commercialization for evaluating the success of the SBIR program. These analysts argue that commercialization is only one of the four overarching SBIR/STTR program goals, and that too strong of a focus on one goal might diminish the emphasis on the others.[66] GAO has noted that using commercialization outcomes as the primary metric of SBIR/STTR success may be insufficient because SBIR and STTR awardees make be making contributions to other agency goals—such as meeting research needs or expanding innovation.[67] Given SBIR/STTR agencies' wide range of missions—from general missions, such as advancing fields of science, to more specific missions, such as providing for the national defense—some analysts have recommended that Congress continue to provide flexibility to agencies in the operation of their programs.[68]

Tracking of Commercialization Successes and Other SBIR Information

Data collection has been and remains an issue for the SBIR and STTR programs according to several reports. An August 2009 GAO study reiterated earlier GAO findings of deficiencies in the SBA Tech-Net system designed to collect information from agency SBIR programs. This report noted that "Although SBA did not meet its statutorily mandated deadline of June 2001, the database has been operational since October 2008, and contains limited new information but may also contain inaccurate historical data."[69] A November 2010 report issued by the SBA's Office of the Inspector General noted that "limited progress" had been made on the Tech-Net system.

> Participating agencies were still experiencing difficulty in searching the database for duplicative awards and other indicators of fraud because information in the Tech-Net database was incomplete, and the search capabilities of the system were limited.... Additionally, SBA had not developed the government-use component of Tech-Net to capture information on the commercialization of SBIR research and development projects.[70]

GAO also addressed agencies' shortcomings with respect to assessing the commercialization success of awardees in reports issued in November 2010 and August 2011. The earlier report found that "DOD lacks complete commercialization data to determine the effectiveness of the program in transitioning space-related technologies into acquisition programs or the commercial sector" and that "there are inconsistencies in recording and defining commercialization."[71] The later study indicated that "Comparable data are not available across participating agencies to evaluate progress in increasing commercialization of SBIR technologies."[72] The report goes on to state that, "with the exception of DOD, agencies that GAO reviewed did not generally take steps to verify commercialization data they collected from award recipients, so the accuracy of the data is largely unknown."[73]

In a December 2013 report, GAO stated that it was "unable to assess the extent of technology transition associated with the military department SBIR programs because comprehensive and reliable technology transition data [for SBIR projects] are not collected."[74] Among the challenges GAO identified in this regard were the lack of a common definition for technology transition across SBIR programs resulting in potential inconsistencies in reporting and the difficulty in tracking transitions due to the long time periods over which they can take place. GAO stated that DOD has not communicated a timeline for when it will be able to comply with statutory reporting requirements. To address these shortcomings, GAO recommended that DOD establish a common definition for technology transition to be used by all DOD SBIR programs; develop a plan to meet new technology transition reporting requirements that will improve the completeness, quality, and reliability of SBIR transition data; and report to Congress on its plan for meeting the reporting requirements set out in P.L. 112-81.[75]

In testimony before the House Small Business Committee in July 2014, GAO once again noted the continuing problem. While acknowledging that DOD agencies have collected selected transition success stories on an ad hoc basis from SBIR program officials, acquisition program officials, prime contractors, and small businesses, GAO found that "the extent of transition is unknown because comprehensive and reliable transition data are not collected."[76] Further, GAO found that the two data systems used by DOD to identify transition successes program-wide "have significant gaps in coverage and data reliability concerns that limit their transition tracking capabilities. In addition, the systems are not designed to capture detailed information on acquisition programs, fielded systems, or on projects that did not transition."[77]

Concerns About Certain Awardees' ("SBIR Shops") Lack of Progress Toward Commercialization

Some critics of the SBIR/STTR programs express particular concern that some firms had become adept at competing for SBIR awards to support their research activities, but had little record of accomplishment in the commercialization of their work. These critics, who sometimes refer to such small businesses as "SBIR shops," assert that these firms may have little interest in commercialization. For example, Lux Research, Inc., an emerging technologies consulting firm, asserts that such firms "go from one SBIR grant to another for years, sometimes decades, and their teams have professional grant writers who are paid to do nothing else but submit successful grant applications into multiple agencies."[78]

Others analysts assert that that while the issue bears watching, the evidence shows that "more of the multiple award winners are also successful in commercialization, receiving additional investment dollars from other sources, and/or successful in having their technologies infused into federal agencies."[79]

Congress responded to such concerns in the 2011 reauthorization act by requiring agencies to track companies success in advancing their work from Phase I to Phase II or from Phase I to Phase III, establishing minimum performance standards in this regard, and denying firms the right to participate in Phase I and Phase II of an agency's SBIR and STTR programs for one year if they fail to meet these standards.

Concerns About Duplicative Awards and Other Types of Waste, Fraud, and Abuse

Identification and elimination of waste, fraud, and abuse in the SBIR/STTR programs have been abiding concerns of Congress. Congress has held multiple hearings, enacted legislation intended to address these concerns, and directed GAO to monitor and report on agency progress in implementing the law and combatting waste, fraud, and abuse. (For example, see "Provisions to Reduce Waste, Fraud, and Abuse" for a discussion of the waste, fraud, and abuse provisions of the 2011 reauthorization act.)

While waste, fraud, and abuse can occur in a variety of ways, duplication of research proposals has been a particular concern for many years. In 1995, GAO reported that contractors had received duplicate funding for similar SBIR research proposals and attributed such duplication to false contractor certifications, lack of a consistent definition for "similar" research," and lack of interagency sharing of data on SBIR awards.[80]

At a 2009 Senate Committee on Commerce, Science, and Transportation hearing, a technology company executive testified that his former employer sought to defraud agency SBIR programs in a number of ways, including duplication in Phase I and Phase II proposals prior to funding; duplication in Phase I and Phase II contracts after funding within performance reports; invoicing the government for the same equipment and materials under different SBIR grants; subcontracting SBIR work out to another company without the government's knowledge; and cross-charging labor and materials used to complete commercial work to government-funded SBIR contracts. The witness further asserted, "To certain types of individuals, the ease that research fraud can be conducted with SBIR funds becomes an addictive alternative to the hard work of commercializing actual research."[81]

At the same hearing, the NASA acting inspector general testified that the agency had investigated or was currently investigating cases of alleged fraud for submitting duplicate proposals to different federal agencies and receiving multiple awards for essentially the same work under the SBIR program; submitting different proposals to multiple federal agencies but providing duplicate deliverables based on the same research; failing to comply with subcontracting limitations; using principal investigators who were not primarily employed by the small business awardee; and failing to perform a substantial portion of the research work contracted by NASA. In addition, he testified that some firms had misrepresented their eligibility, including false assertions of American ownership and meeting the small business size standard. While testifying that NASA had taken corrective measures to address vulnerabilities to waste, fraud, and abuse, he noted that "in the cases that we are conducting today, we still see the same violations that we saw as early as 1992."[82]

Among its provisions, the 2011 reauthorization act directs the SBA to amend its SBIR and STTR policy directives to include definitions or descriptions of fraud, waste, and abuse. The amended directives now identify a variety of actions that constitute waste, fraud, or abuse, including:

- misrepresentations or material, factual omissions to obtain, or otherwise receive funding under, an SBIR award;
- misrepresentations of the use of funds expended, work done, results achieved, or compliance with program requirements under an SBIR award;
- misuse or conversion of SBIR award funds, including any use of award funds while not in full compliance with SBIR program requirements, or failure to pay taxes due on misused or converted SBIR award funds;
- fabrication, falsification, or plagiarism in applying for, carrying out, or reporting results from an SBIR award;
- failure to comply with applicable federal costs principles governing an award;
- extravagant, careless, or needless spending;
- self-dealing, such as making a sub-award to an entity in which the principal investigator has a financial interest;
- acceptance by agency personnel of bribes or gifts in exchange for grant or contract awards or other conflicts of interest that prevents the government from getting the best value; and
- lack of monitoring, or follow-up if questions arise, by agency personnel to ensure that awardee meets all required eligibility requirements, provides all required certifications, performs in

accordance with the terms and conditions of the award, and performs all work proposed in the application.[83]

The 2011 authorization act required GAO to publish an initial report within one year from the date of enactment of the act and every four years thereafter on agency efforts to combat waste, fraud, and abuse and comply with the provisions of the act in this regard. In November 2012, GAO published *Small Business Research Programs: Agencies Are Implementing New Fraud, Waste, and Abuse Requirements*. The GAO report found that the SBA had revised its SBIR and STTR policy directives in August 2012 to include new requirements to help agencies identify and prevent waste, fraud, and abuse, including 10 minimum requirements that all SBIR/STTR agencies must meet. GAO also found that while SBIR and STTR programs varied in their plans to implement the new requirements, program managers did not anticipate significant challenges in this regard. GAO also noted that each agency already had in place tools to address or partially address the new requirements.

SBA Delays in Meeting Statutory Reporting Requirements

The Small Business Act has required the SBA to report annually to Congress on the SBIR and STTR programs since the inception of these programs. SBA compliance with this requirement has been an ongoing issue. Most recently, the SBA did not produce an annual report to Congress for FY2009 or FY2010, instead producing a single report covering the three-year period from FY2009 to FY2011. In addition, as of July 25, 2014, the SBA had not delivered an FY2012 or an FY2013 annual report to Congress. Failure to produce these reports on a timely basis may impede Congress's exercise of its oversight responsibilities. Among the issues that may affect the timeliness of SBA reporting are SBIR/STTR agencies' delays in providing data to the SBA and adequate staffing levels at SBA devoted to producing the report.

Other Issues

As the 2011 reauthorization law is implemented, Congress may decide to explore how the new provisions affect program operation and outcomes including efforts to identify and eliminate duplication of awards and to protect the rights of small businesses to data generated in the performance of an SBIR award. In addition, some experts question whether the SBIR and STTR programs are meeting their different mandated objectives. Other critics maintain that the government has no role in directly supporting industrial research and development. These and other issues may be debated as the SBIR and STTR programs continue to function through September 30, 2017.

End Notes

1 Small Business Innovation Development Act of 1982 (P.L. 97-219).

2 Ibid.

3 Small Business Research and Development Enhancement Act of 1992 (P.L. 102-564).

4 The SBIR/STTR Reauthorization Act of 2011 was enacted as Division E of the National Defense Authorization Act for Fiscal Year 2012.

5 The percentages identified in law that must be set aside for SBIR and STTR are minimums; agencies may opt to set aside more than these percentages.

6 Email communication between SBA and CRS on March 26, 2014. The same FY2011 data are included as a table in the SBA's Small Business Innovation Research (SBIR) and Small Business Technology Transfer (STTR) Program Annual Report FY2009-FY2011.

7 Telephone communication between SBA and CRS on July 18, 2014.

8 SBIR/STTR Reauthorization Act of 2011, enacted as Division E of the National Defense Authorization Act for Fiscal Year 2012 (P.L. 112-81).

9 U.S. Government Accountability Office, Small Business Research Programs: More Guidance and Oversight Needed to Comply with Spending and Reporting Requirements, GAO-14-431, June 2014.

10 The SBA directive is required under Section 9(j) of the Small Business Act (15 U.S.C. §638). The directive was updated on February 24, 2014. The directive is available at http://sbir.gov/sites/default/files/ sbir_sttr_program_overview_tips_for_applicants.pdf.

11 See, for example, U.S. Government Accountability Office, Small Business Research Programs: Agencies Are Implementing New Fraud, Waste, and Abuse Requirements, GAO-13-70R, November 15, 2012, p. 1, http://gao.gov/ assets/660/650129.pdf.

12 15 U.S.C. §638.

13 The STTR program is discussed in more detail later in this report.

14 Small Business Administration, Small Business Innovative Research Program Policy Directive, February 24, 2014.

15 15 U.S.C. §638(e)(4)(b).

16 P.L. 112-81, §5108.

17 P.L. 112-81, §5121.

18 15 U.S.C. §638(q)(1).

19 See 13 C.F.R. §121.701.

20 According to SBA, "The exception to this is if the VC is itself more than 50% directly owned and controlled by one or more individuals who are citizens or permanent resident aliens of the United States. In such a case, that VC is allowed to have majority ownership and control of the awardee. In that case, the VC and the awardee, and all other affiliates, must have a

total of 500 employees or less." Source: SBIR/STTR website, http://sbir.gov/faq/vc-participation.

[21] 13 C.F.R. §121.702.

[22] Small Business Administration, Small Business Innovation Research Program Policy Directive, February 24, 2014.

[23] See "Data Sources and Limitations" above.

[24] The success rate is the number of successful proposals divided by total proposals submitted, expressed as a percentage.

[25] Phase II funding includes original and subsequent Phase II award funding, as well as modifications.

[26] According to the SBA,

Some minority groups are presumed to be socially and economically disadvantaged and can qualify for the 8(a) program. [See description of the SBA 8(a) Business Development Program below.] These groups include: African Americans, Hispanic Americans, Native Americans, Asian Pacific Americans and Subcontinent Asian Americans. Individuals who are not members of one or more of these groups can be considered for the 8(a) program, but they must provide substantial evidence and documentation that demonstrates that they have been subjected to bias or discrimination and are economically disadvantaged. Firms owned by Alaska Native Corporations, Indian Tribes, Native Hawaiian Organizations and Community Development Corporations can also apply to the program.

For additional information on minority and disadvantaged businesses, see the "Minority-Owned Businesses" webpage on the SBA's website at http://www.sba.gov/content/minority-owned-businesses.

The SBA's 8(a) Business Development Program is "a business assistance program for small disadvantaged businesses. The 8(a) Program offers a broad scope of assistance to firms that are owned and controlled at least 51% by socially and economically disadvantaged individuals." For additional information, see the "8(a) Business Development Program" webpage on the SBA's website at http://www.sba.gov/content/about-8a-business-development-program.

[27] According to the SBA, HUBZones are designated by statute and draw upon determinations and information obtained by other agencies. A HUBZone may be one of the following: a qualified Census Tract, a qualified Nonmetropolitan County, a qualified Indian reservation, a qualified Base Closure Area, or a redesignated area. For more information, see the SBA's Understanding HUBZone Designations webpage at http://www.sba.gov/tools/sba-learning-center/training/ hubzone-mini-primer-understanding-hubzone-designations.

[28] National Science Foundation, Federal Funds for Research and Development: Fiscal Years 2010–12, NSF 13-326, 2013, Table 122, http://www.nsf.gov/statistics/nsf13326/content.cfm?pub_id=4243&id=2.

[29] According to the SBA, an eligible "research institution" is defined, for purposes of the STTR, as:

one that has a place of business located in the United States, which operates primarily within the United States or which makes a significant contribution to the U.S. economy through payment of taxes or use of American products, materials or labor, and is: (1) A non-profit institution as defined in section 4(5) of the Stevenson-Wydler Technology Innovation Act of 1980 (that is, an organization that is owned and operated exclusively for scientific or educational purposes, no part of the net earnings of which inures to the benefit of any private shareholder or individual) and includes non-profit medical and surgical hospitals; or (2) A federally-funded R&D center as identified by the National Science Foundation in accordance with the Government-wide Federal Acquisition Regulation issued in accordance with section 35(c)(1) of the Office of Federal Procurement Policy Act (or any successor regulation thereto).

Source: Small Business Administration, Small Business Technology Transfer Policy Directive (Updated February 24, 2014).

[30] Small Business Administration, SBIR/STTR website, "About STTR," http://www.sbir.gov /about/about-sttr.

[31] The SBA directive is required under Section 9(j) of the Small Business Act (15 U.S.C. §638). The directive was updated on February 24, 2014. The directive is available at http://sbir. gov/sites/default/files/ sbir_sttr_program_overview_tips_for_applicants.pdf.

[32] See, for example, U.S. Government Accountability Office, Small Business Research Programs: Agencies Are Implementing New Fraud, Waste, and Abuse Requirements, GAO-13-70R, November 15, 2012, p. 1, http://gao.gov/ assets/660/650129.pdf.

[33] Small Business Administration, Small Business Technology Transfer Program Policy Directive, February 24, 2014.

[34] A federal agency may, however, issue an STTR Phase II award to an SBIR Phase I awardee to further develop the work performed under the SBIR Phase I award. An agency -

must base its decision upon the results of the work performed under the Phase I award and the scientific and technical merit, and commercial potential of the Phase II proposal. The SBIR Phase I awardee must meet the eligibility and program requirements of the STTR Program in order to receive the STTR Phase II award.

Source: Small Business Administration, Small Business Technology Transfer Program Policy Directive, February 24, 2014, p.11.

[35] P.L. 112-81, §5108.

[36] P.L. 112-81 §5121.

[38] 13 C.F.R. §121.702.

[39] SBIR/STTR website, Frequently Asked Questions—VC Participation, http://sbir.gov/faq/vc-participation.

[40] Small Business Administration, Small Business Technology Transfer Program Policy Directive, February 24, 2014, p.15.

[41] As defined in 15 U.S.C. §3703(3) a nonprofit institution is "an organization owned and operated exclusively for scientific or educational purposes, no part of the net earnings of which inures to the benefit of any private shareholder or individual."

[42] Chapter 35 of the Federal Acquisition Regulation provides the following explanation and purposes of FFRDCs:

An FFRDC meets some special long-term research or development need which cannot be met as effectively by existing in-house or contractor resources. FFRDC's enable agencies to use private sector resources to accomplish tasks that are integral to the mission and operation of the sponsoring agency.... FFRDC's are operated, managed, and/or administered by either a university or consortium of universities, other not-for-profit or nonprofit organization, or an industrial firm, as an autonomous organization or as an identifiable separate operating unit of a parent organization.

A list of FFRDCs is maintained by the National Science Foundation. See http://www.nsf.gov/statistics/ffrdclist/.

[43] See "Data Sources and Limitations," above.

[44] See footnote 26 for a description of the SBA 8(a) Business Development Program and additional information on minority and disadvantaged businesses.

[45] According to the SBA, HUBZones are designated by statute and draw upon determinations and information obtained by other agencies. A HUBZone may be one of the following: a qualified Census Tract, a qualified Nonmetropolitan County, a qualified Indian Reservation, a Qualified Base Closure Area, or a Redesignated Area. For more information, see the SBA's Understanding HUBZone Designations webpage at http://www.sba.gov/tools/sba-learning-center/ training/hubzone-mini-primer-understanding-hubzone-designations.

[46] P.L. 112-81 §5107.

[47] U.S. Government Accountability Office, Small Business Research Programs: Actions Needed to Improve Compliance with Spending and Reporting Requirements, GAO-13-421, September 9, 2013, http://gao.gov/assets/660/ 657489.pdf.

[48] P.L. 112-81 §5111.

[49] P.L. 112-81 §5123.

[50] 15 U.S.C. §638.

[51] U.S. Government Accountability Office, Small Business Innovation Research: Data Rights Protections, GAO-14- 116R, November 4, 2013, http://gao.gov/assets/660/658721.pdf.

[52] Small Business Administration, "Small Business Size Regulations, Small Business Innovation Research (SBIR) Program and Small Business Technology Transfer (STTR) Program," 72 Federal Register 76215, December 27, 2012.

[53] U.S. Government Accountability Office, Small Business Research Programs: Actions Needed to Improve Compliance with Spending and Reporting Requirements, GAO-13-421, September 9, 2013, http://gao.gov/assets/660/ 657489.pdf.

[54] U.S. Government Accountability Office, Small Business Research Programs: More Guidance Needed to Comply with Spending and Reporting Requirements, GAO-14-431, summary page, June 2014, http://gao.gov/assets/670/ 663909.pdf.

[55] "Extramural budget" is defined as "the sum of the total obligations for R/R&D minus amounts obligated for R/R&D activities by employees of a federal agency in or through government-owned, government-operated facilities. See SBIR Program Policy Directive, p. 6.

[56] For example, the Department of Energy extramural R/R&D budget includes funding in the Office of Science, Office of Nuclear Energy, Office of Electricity Delivery and Energy Reliability, Office of Energy Efficiency and Renewable Energy, Office of Environmental Management, Office of Fossil Energy, National Nuclear Security Administration, and Advanced Research Projects Agency—Energy.

[57] For example, more than one National Science Foundation account has extramural R&D funding as well as funding that is not R&D.

[58] Agencies are required to exclude, for example, subunits in the intelligence community from their extramural R&D budget.

[59] 15 U.S.C. §638(i)(2).

[60] Small Business Administration, Small Business Innovative Research Program Policy Directive, February 24, 2014, p. 40.

[61] U.S. Government Accountability Office, Small Business Research Programs: More Guidance and Oversight Needed to Comply with Spending and Reporting Requirements, GAO-14-431, June 2014.

[62] 15 U.S.C. §638(e).

[63] The U.S. Government Accountability Office defines obligation as:

> A definite commitment that creates a legal liability of the government for the payment of goods and services ordered or received, or a legal duty on the part of the United States that could mature into a legal liability by virtue of actions on the part of the other party beyond the control of the United States. Payment may be made immediately or in the future. An

agency incurs an obligation, for example, when it places an order, signs a contract, awards a grant, purchases a service, or takes other actions that require the government to make payments to the public or from one government account to another.

Source: U.S. Government Accountability Office, A Glossary of Terms Used in the Federal Budget Process, GAO-05- 734SP, September 2005, p. 70, http://gao.gov/ assets/80/ 76911. pdf.

[64] U.S. Government Accountability Office, Small Business Research Programs: More Guidance and Oversight Needed to Comply with Spending and Reporting Requirements, GAO-14-431, June 2014.

[65] U.S. Government Accountability Office, Small Business Research Programs: More Guidance and Oversight Needed to Comply with Spending and Reporting Requirements, GAO-14-431, June 2014.

[66] Testimony of David H. Finifter, Professor of Economics, Emeritus, Research Professor of Public Policy, The College of William and Mary, in U.S. Congress, House Committee on Small Business, Oversight of the Small Business Innovation Research and Small Business Technology Transfer Programs, hearings, 113th Cong., 2nd sess., May 21, 2014, available at http://smallbusiness.house.gov/calendar/eventsingle.aspx?EventID=373098.

[67] U.S. Government Accountability Office, Federal Research: Observations on the Small Business Innovation Research Program, GAO-05-861T, June 28, 2005, http://gao.gov/ assets/120/111851.pdf.

[68] Testimony of David H. Finifter, Professor of Economics, Emeritus, Research Professor of Public Policy, The College of William and Mary, in U.S. Congress, House Committee on Small Business, Oversight of the Small Business Innovation Research and Small Business Technology Transfer Programs, hearings, 113th Cong., 2nd sess., May 21, 2014, available at http://smallbusiness.house.gov/calendar/eventsingle.aspx?EventID=373098.

[69] Government Accountability Office, Small Business Innovation Research: Observations on Agencies' Data Collection and Eligibility Determination Efforts, GAO-09-956T, August 6, 2009, p. 11, http://www.gao.gov/products/GAO-09- 956T.

[70] Small Business Administration, Office of the Inspector General, Usefulness of the Small Business Innovation Research Tech-Net Database, Report Number 11-02, November 12, 2010, p. 3, http://www.sba.gov/sites/default/files/ oig_report_11_02.pdf.

[71] U.S. Government Accountability Office, Space Acquisitions: Challenges in Commercializing Technologies Developed under the Small Business Innovation Research Program, GAO-11-21, November 10, 2010, summary page, http://gao.gov/assets/320/312130.pdf.

[72] U.S. Government Accountability Office, Small Business Innovation Research: SBA Should Work with Agencies to Improve the Data Available for Program Evaluation, GAO-11-698, August 15, 2011, summary page, http://gao.gov/ assets/330/322653.pdf.

[73] U.S. Government Accountability Office, Small Business Innovation Research: SBA Should Work with Agencies to Improve the Data Available for Program Evaluation, GAO-11-698, August 15, 2011, summary page, http://gao.gov/ assets/330/322653.pdf.

[74] U.S. Government Accountability Office, Small Business Innovation Research: DOD's Program Supports Weapon Systems Development, but Lacks Comprehensive Data on Technology Transition Outcomes, GAO-14-96, December 2013, p. 8, http://www.gao. gov/assets/660/659874.pdf.

[75] Ibid.

[76] U.S. Government Accountability Office, Small Business Innovation Research: DOD's Program Has Developed Some Technologies that Support Military Users, but Lacks

Comprehensive Data on Transition Outcomes, GAO-14- 748T, July 23, 2014, p. 6, http://gao.gov/assets/670/664971.pdf.

77 Ibid, p. 6.

78 Bilal Zuberi, Partner, Lux Research, Inc., "SBIR/STTR grants are great. 'SBIR shops' are not," May 20, 2014, http://www.luxcapital.com/blog/sbirsttr-grants-are-great-sbir-shops-are-not/.

79 Testimony of David H. Finifter, Professor of Economics, Emeritus, Research Professor of Public Policy, The College of William and Mary, in U.S. Congress, House Committee on Small Business, Oversight of the Small Business Innovation Research and Small Business Technology Transfer Programs, hearings, 113th Cong., 2nd sess., May 21, 2014, available at http://smbiz.house.gov/UploadedFiles/5-21-2014_Finifter_Testimony.pdf.

80 U.S. General Accounting Office, Federal Research: Interim Report on the Small Business Innovation Research Program, GAO/RCED 95-59, March 8, 1995, http://gao. gov/ assets/ 160/154923.pdf.

81 Testimony of Alfred J. Longhi, Jr., former Vice President, Lithium Power Technologies, in U.S. Congress, Senate Committee on Commerce, Science, and Transportation, Waste, Fraud, and Abuse in the SBIR Program, hearings, 112th Cong., 1st sess., August 6, 2009, available at http://www.gpo.gov/fdsys/pkg/CHRG-111shrg52753/html/CHRG111shrg 52753.htm.

82 Testimony of Thomas J. Howard, Acting Inspector General, NASA, in U.S. Congress, Senate Committee on Commerce, Science, and Transportation, Waste, Fraud, and Abuse in the SBIR Program, hearings, 112th Cong., 1st sess., August 6, 2009, available at http://www. gpo.gov/fdsys/pkg/CHRG-111shrg52753/html/CHRG111shrg52753.htm.

83 Small Business Administration, Small Business Innovative Research Program Policy Directive, February 24, 2014.

In: Small Business Innovation Research … ISBN: 978-1-63482-532-0
Editor: Thomas H. Chavez

Chapter 2

SMALL BUSINESS RESEARCH PROGRAMS: MORE GUIDANCE AND OVERSIGHT NEEDED TO COMPLY WITH SPENDING AND REPORTING REQUIREMENTS*

United States Government Accountability Office

WHY GAO DID THIS STUDY

Federal agencies have awarded more than 156,000 contracts and grants, totaling nearly $40 billion, through the SBIR and STTR programs to small businesses to develop and commercialize innovative technologies. The Small Business Act requires agencies with extramural R&D budgets that meet certain thresholds for participation—$100 million for SBIR and $1 billion for STTR—must spend a percentage of these annual budgets on the programs. The agencies are to report on their activities to SBA and, in turn, SBA is to report to Congress.

The 2011 reauthorization of the programs mandated GAO to review compliance with spending and reporting requirements, and other program aspects. This report addresses, for fiscal year 2012, (1) the extent to which agencies complied with spending requirements, (2) the extent to which

* This is an edited, reformatted and augmented version of a United States Government Accountability Office publication, No. GAO-14-431, dated June 2014.

agencies and SBA complied with certain reporting requirements, (3) the potential effects of basing spending requirements on agencies' total R&D budgets, and (4) what is known about the amounts spent administering the programs. GAO reviewed agency spending data and required reports for fiscal year 2012 and interviewed program officials from SBA and the participating agencies.

What GAO Recommends

GAO recommends, among other things, that SBA revise program policy directives to accurately summarize spending requirements and request that agencies submit their methodology reports on time. SBA and participating agencies generally agreed with GAO's findings and recommendations.

What GAO Found

Agency data indicate that 8 of the 11 agencies participating in the Small Business Innovation Research (SBIR) program and 2 of the 5 agencies participating in the Small Business Technology Transfer (STTR) program complied with spending requirements in fiscal year 2012. Program managers for agencies that did not comply with the requirements identified reasons for noncompliance. For example, program managers at two of the agencies told GAO that they believe their agencies comply with spending requirements if the agencies spend the total amount reserved or budgeted for the program, regardless of the year the funding is spent. However, the authorizing legislation for the programs requires agencies to "expend" a certain amount of funding each year. This difference in the interpretation of spending requirements occurred, in part, because the Small Business Administration's (SBA) policy directives for the programs inaccurately state that the authorizing legislation requires agencies to "reserve" the minimum amount each year. Additionally, some officials told GAO their agencies did not comply with spending requirements because the recent reauthorization of the programs included an increased spending requirement in fiscal year 2012, but the reauthorization was enacted a full quarter into the fiscal year, after some agencies had planned their programs and made awards.

Participating agencies and SBA did not fully comply with certain reporting requirements for the SBIR and STTR programs. For example, participating agencies are required to submit reports to SBA describing their methodologies for calculating their budgets for extramural research or research and development (R&D)—which is generally conducted by nonfederal employees outside of federal facilities—within 4 months of the enactment of appropriations. However, all 11 participating agencies were late in submitting these reports because SBA allowed them to submit the reports later. As a result, SBA was unable to analyze the reports and provide timely feedback to assist agencies in accurately calculating these budgets.

Potential effects of basing each participating agency's spending requirement on its total R&D budget instead of its extramural R&D budget include an increase in the amount of the spending requirement—for some agencies more than others— and, if the thresholds for participation in the programs did not change, an increase in the number of agencies required to participate. Officials identified benefits of such a change, such as funding more projects, but they generally said the drawbacks could outweigh the benefits.

Little is known about the total amounts agencies spent administering the SBIR and STTR programs because agencies did not consistently collect such information for fiscal year 2012. Agencies are not required to track costs for administering the programs. Most agencies provided GAO with some data on such costs for fiscal year 2012, ranging from about $200,000 to about $8 million, but the data were wide-ranging, incomplete, and unverifiable. With the start of a pilot program in fiscal year 2013 that allows agencies to use up to 3 percent of SBIR program funds for administrative costs, agencies will be required to report to SBA on the amount spent for such activities. However, even with the pilot program, agencies will likely not identify or track all administrative costs.

ABBREVIATIONS

R&D	research or research and development
SBA	Small Business Administration
SBIR	Small Business Innovation Research
STTR	Small Business Technology Transfer

June 6, 2014

Congressional Committees

Since the early 1980s, federal agencies have awarded more than 156,000 contracts and grants totaling nearly $40 billion through two small business research programs: the Small Business Innovation Research (SBIR) and Small Business Technology Transfer (STTR) programs. The SBIR and STTR programs are similar in that participating agencies identify topics for research or research and development (R&D) projects and make awards to small businesses to develop and commercialize innovative technologies. However, for the SBIR program, a for-profit small business performs the work associated with an award, while the STTR program requires the small business to partner with a nonprofit research institution—such as a nonprofit college or university or federally funded research and development center—to perform the work. Federal agencies with a budget of $100 million or more for extramural R&D are required to establish and operate an SBIR program, while federal agencies with budgets of $1 billion or more for extramural R&D are required to establish and operate an STTR program.[1] Currently, 11 agencies participate in the SBIR program, and 5 of these agencies also participate in the STTR program, as shown in table 1.

Table 1. Agencies Participating in the Small Business Innovation Research (SBIR) and Small Business Technology Transfer (STTR) Programs

Agency	Program	
	SBIR	STTR
Department of Agriculture (USDA)	X	
Department of Commerce	X	
Department of Defense (DOD)	X	X
Department of Education	X	
Department of Energy (DOE)	X	X
Department of Health and Human Services (HHS)	X	X
Department of Homeland Security (DHS)	X	
Department of Transportation (DOT)	X	
Environmental Protection Agency (EPA)	X	
National Aeronautics and Space Administration (NASA)	X	X
National Science Foundation (NSF)	X	X

Source: Small Business Administration.

The Small Business Act, which authorizes the programs, establishes the minimum percentage of an agency's extramural R&D budget that must be spent on the programs annually.[2] In fiscal year 2012, agencies participating in the SBIR program were required to spend at least 2.6 percent of their extramural R&D budgets on the program and agencies participating in the STTR program were required to spend at least 0.35 percent of their extramural R&D budgets on the program. The SBIR and STTR policy directives require participating agencies to submit data to the Small Business Administration (SBA) each year on the amount of their extramural R&D budgets and the amount obligated for awards, among other information. The Small Business Act also establishes certain reporting requirements for participating agencies and SBA. Among other things, agencies are to report to SBA on their methodologies for calculating their extramural R&D budgets within 4 months of the enactment of their annual appropriations. Furthermore, SBA is to annually report to Congress on all participating agencies' SBIR and STTR programs.

In addition, the 2011 reauthorization of the programs mandates that GAO review the participating agencies' compliance with spending and reporting requirements for the programs, as well as other aspects of the programs.[3] We issued our first annual report in response to this mandate in September 2013, which covered fiscal years 2006 through 2011.[4] In that report, we recommended, among other things, that SBA provide more guidance to agencies for calculating spending requirements under certain circumstances and provide more timely reports to Congress. This second annual report determines for fiscal year 2012 (1) the extent to which participating agencies complied with the programs' spending requirements, (2) the extent to which participating agencies and SBA complied with certain reporting requirements, (3) the potential effects of basing the spending requirements for the SBIR and STTR programs on agencies' total R&D budgets instead of their extramural R&D budgets, and (4) what is known about the amounts participating agencies spent for administering the programs.

To address these objectives, we followed the methodology that we used for our first annual report on these issues. To determine the extent to which participating agencies complied with the programs' spending requirements in fiscal year 2012, we used the data that each agency submitted to SBA and calculated each agency's spending requirement by applying the mandated percentages to the agency's extramural R&D budget. We then compared the spending requirements to the total obligations data the agencies submitted to SBA for fiscal year 2012.[5] We determined that an agency complied with its spending requirement if the agency's spending for these programs was equal to or greater than the calculated spending requirement.[6] To assess the reliability of these data, we interviewed agency officials about the source of the data and data quality control procedures for the data, reviewed relevant documentation, and compared the agency's submissions to SBA with SBA's version of documents received from the agencies. We determined the data were sufficiently reliable for the purposes of this report. We discussed agencies' compliance with spending requirements, including reasons agencies did not meet spending requirements, with program officials at each of the participating agencies.

To determine the extent to which participating agencies and SBA complied with certain reporting requirements for fiscal year 2012, we compared information in the methodology reports that each agency submitted to SBA to requirements in the law and SBA policy directives. Additionally, we requested SBA's report to Congress for fiscal year 2012 and discussed the status of that report with SBA program officials, who said it had not been submitted at the time of our review.

To determine the potential effects of basing spending requirements for the SBIR and STTR programs on agencies' total R&D budgets instead of their extramural R&D budgets, we used fiscal year 2012 data on total R&D budget authority from the President's budget to calculate potential spending requirements for each federal agency.[7] In calculating these potential spending requirements, we assumed that the same spending percentages currently required by the Small Business Act would apply to total R&D budgets. We also assumed that the current spending thresholds that require agency participation in the SBIR and STTR programs when applied to extramural R&D budgets would apply to total R&D budgets. In other words, an agency with a total annual R&D budget of $100 million or more would be required to participate in the SBIR program, while an agency with a total annual R&D budget of $1 billion or more would be required to participate in both the SBIR and STTR programs. We compared the spending requirements that would result under this scenario with the spending requirements under the current law to determine the potential effects that changing the methodology to calculate the spending requirements could have had for fiscal year 2012, recognizing that changes to the thresholds for participating in the programs or the mandated percentages could have different effects. We also calculated spending requirements under another scenario to determine whether different assumptions would lead to different outcomes. Under this scenario, we applied a lower percentage than is required under current law to each agency's total

R&D budget. Additionally, we discussed the effects of changing the methodology for calculating the spending requirements for the programs with program officials at each of the agencies currently participating in the programs and offered the agencies that could be affected by the change an opportunity to comment on our analysis.

To determine what is known about the amounts participating agencies spent for administering the programs, we collected administrative cost data from agencies and discussed the data with program officials. We determined that these data were too incomplete and from such varied sources that an assessment of the available data's reliability was not possible. Consequently, we could not use the data to draw conclusions about agencies' administrative costs in our report.

We conducted this performance audit from November 2013 to June 2014 in accordance with generally accepted government auditing standards. Those standards require that we plan and perform the audit to obtain sufficient, appropriate evidence to provide a reasonable basis for our findings and conclusions based on our audit objectives. We believe that the evidence obtained provides a reasonable basis for our findings and conclusions based on our audit objectives.

BACKGROUND

The SBIR program was initiated in 1982 and has four main purposes: (1) use small businesses to meet federal R&D needs, (2) stimulate technological innovation, (3) increase commercialization of innovations derived from federal R&D efforts, and (4) encourage participation in technological innovation by small businesses owned by women and disadvantaged individuals. The purpose of the STTR program—initiated in 1992—is to stimulate a partnership of ideas and technologies between innovative small businesses and research institutions through federally funded R&D.[8] The National Defense Authorization Act for Fiscal Year 2012, enacted on December 31, 2011, reauthorized the SBIR and STTR programs through September 30, 2017.[9]

The SBIR and STTR programs each include the following three phases:

- In phase I, agencies make awards to small businesses to determine the scientific and technical merit and feasibility of ideas that appear to have commercial potential. Phase I awards normally do not exceed

$150,000. For SBIR, phase I awards generally last 6 to 9 months and for STTR these awards generally last 1 year.

- In phase II, small businesses with phase I projects that demonstrate scientific and technical merit and feasibility, in addition to commercial potential, may compete for awards of up to $1 million to continue the R&D for an additional period, normally not to exceed 2 years.
- Phase III is for small businesses to pursue commercialization of technology developed in prior phases. Phase III work derives from, extends, or completes an effort made under prior phases, but it is funded by sources other than the SBIR or STTR programs. In this phase, small businesses are expected to raise additional funds from private investors, the capital markets, or from funding sources within the agency that made the initial award other than its SBIR or STTR program. While SBIR or STTR funding cannot be used for phase III, agencies can participate in phase III by, for example, purchasing the technology developed in prior phases.[10]

SBA's Office of Investment and Innovation is responsible for overseeing and coordinating the participating agencies' efforts for the SBIR and STTR programs by setting overarching policy and issuing policy directives, collecting program data, reviewing agency progress, and reporting annually to Congress, among other responsibilities. As part of its oversight and coordination role, SBA issued SBIR and STTR policy directives in September 2002 and December 2005, respectively, and updated them in August 2012, January 2014, and February 2014. These directives explain and outline additional requirements for agencies' implementation of the SBIR and STTR programs. Among the topics discussed, the policy directives include a listing of the data that agencies are to submit to SBA annually—such as their extramural R&D budget amounts and the amounts obligated for awards for the programs—and call for agencies to submit reports describing their methodologies for calculating their extramural R&D budgets within 4 months of the enactment of appropriations.

Each participating agency must manage its SBIR and STTR programs in accordance with program laws, regulations, and the policy directives issued by SBA. In general, the programs are similar across agencies. For those agencies that have both SBIR and STTR programs, agencies usually use the same process for both programs. Each participating agency has considerable flexibility to design and manage the specifics of these programs, such as determining research topics, selecting award recipients, and administering funding agreements. All of the agencies follow the same general process to obtain proposals from and make awards to small businesses for both the SBIR and STTR programs. At least annually, each participating agency issues a solicitation requesting proposals for projects in topic areas determined by the agency. Each agency uses its own process to review proposals and determine which proposals should receive awards. Also, each agency determines whether the funding for awards will be provided as grants or contracts.

DATA INDICATE NOT ALL AGENCIES MET SPENDING REQUIREMENTS

Agency data indicate that not all agencies complied with the SBIR and STTR spending requirements for fiscal year 2012. Program managers identified several reasons why agencies did not comply with the spending requirements.

Agency Data Indicate Most Agencies Complied with SBIR Spending Requirements

The data agencies submitted to SBA indicates that 8 of the 11 participating agencies met or exceeded their fiscal year 2012 spending requirements for the SBIR program, while 3 of the agencies did not meet the requirements (see fig. 1 and app. I for additional data).[11] According to the agencies' data, the 8 agencies that met or exceeded the requirements spent between 2.6 percent and 4.0 percent of their extramural R&D budgets on the program, and the remaining 3 agencies spent between 2.3 percent and 2.5 percent. In comparison, agency data indicated that 10 of the 11 agencies met or exceeded spending requirements in fiscal year 2011, and that 3 of the 11 agencies complied with spending requirements each fiscal year from 2006 through 2011, as we found in September 2013.[12]

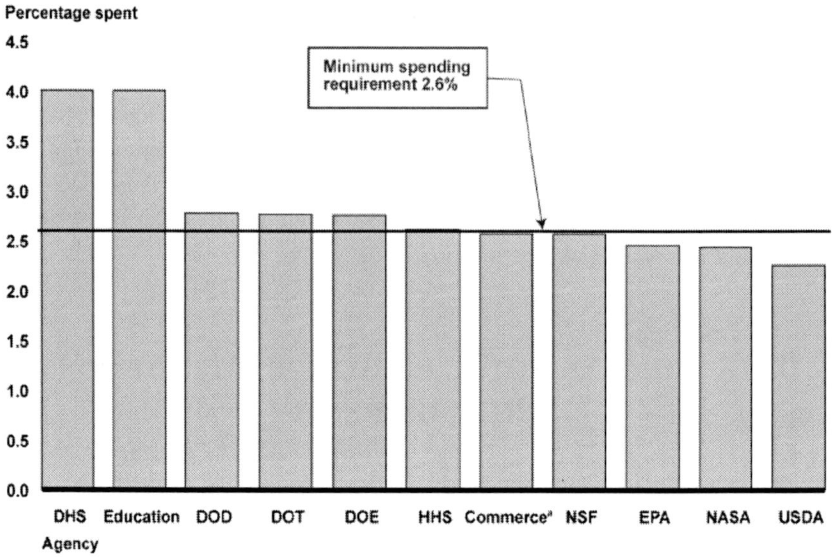

Source: GAO analysis of agency data submitted to SBA. | GAO-14-431.

Note: We defined compliance as spending at least 2.6 percent of an agency's extramural research or research and development (R&D) budget on the Small Business Innovation Research (SBIR) program in fiscal year 2012, as required by the Small Business Act. This method is consistent with SBA's approach for determining compliance with spending requirements in its reports to Congress on the program. We used the agencies' obligations data to represent spending for the programs, in part because obligations data were readily available from each of the agencies for program purposes, and because obligations provided a reasonable measure of the spending for the programs in each year.

[a]Officials from one subunit within the Department of Commerce—the National Institute of Standards and Technology—said that they calculate their spending requirement for the program based on their actual obligations for extramural R&D for the previous fiscal year, rather than their extramural R&D budget for the current fiscal year, but they do not report this figure to SBA. We used the data that the Department of Commerce submitted to SBA to calculate its spending requirements.

Figure 1. Percentage of Extramural Research or Research and Development Budgets That Participating Agencies Spent for the Small Business Innovation Research Program in Fiscal Year 2012.

Agency Data Indicate Two Agencies Complied with STTR Spending Requirements

Our analysis of the data agencies submitted to SBA indicates that two of the five agencies participating in the STTR program met or exceeded their fiscal year 2012 spending requirements, while the remaining three agencies did not meet the requirements (see fig. 2 and app. II for additional data).[13] According to the agencies' data, the two agencies that complied with the requirements spent between 0.35 percent and 0.38 percent of their extramural R&D budgets on their STTR programs, and the three agencies that did not comply spent between 0.30 percent and 0.32 percent. In comparison, the data that agencies submitted to SBA indicated that two of the five agencies, HHS and NASA, complied with spending requirements in fiscal year 2011, and one of the agencies— HHS—complied with spending requirements each fiscal year from 2006 to 2011, as we found in September 2013.[14]

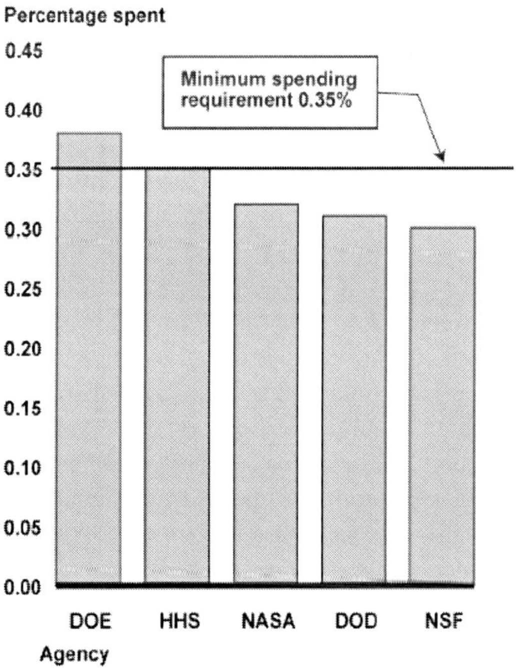

Source: GAO analysis of agency data submitted to SBA. | GAO-14-431.

Note: We defined compliance as spending at least 0.35 percent of an agency's extramural R&D budget on the STTR program in fiscal year 2012, as required by the Small Business Act.

This method is consistent with SBA's approach for determining compliance with spending requirements in its reports to Congress on the program. We used the agencies' obligations data to represent spending for the programs, in part because obligations data were readily available from each of the agencies for program purposes, and because obligations provided a reasonable measure of the spending for the programs in each year.

Figure 2. Percentage of Extramural Research or Research and Development Budgets That Participating Agencies Spent for the Small Business Technology Transfer Program in Fiscal Year 2012.

Program Managers Identified Reasons Agencies Did Not Comply with Spending Requirements

Program managers at the agencies that did not comply with spending requirements for fiscal year 2012 identified reasons that the agencies did not comply: (1) some agencies reserved funds to spend in a subsequent fiscal year without spending the minimum required amount in fiscal year 2012; (2) the spending requirement increased part of the way through the year, after agencies made spending plans; (3) agencies did not submit data on all of their spending; and (4) the amount agencies spent on extramural R&D was higher than estimated at the beginning of the fiscal year. Some of these circumstances also applied to agencies that complied with spending requirements in fiscal year 2012.

Some Agencies Reserved Funds to Spend in a Subsequent Fiscal Year

Two of the agencies that did not comply with spending requirements—DOD and USDA—reserved funds to spend in a subsequent fiscal year without spending the minimum required amount in fiscal year 2012.[15] The Small Business Act requires participating agencies to "expend" the minimum percentages of their extramural R&D budgets with small businesses annually. Additionally, SBA officials said they expect agencies to spend at least the minimum required percentages on the programs each year. However, program officials at DOD and USDA told us that they believe their agencies comply with spending requirements if the agencies spend the total amount reserved or budgeted for the program, regardless of the year the funding is spent. We found that these two agencies did not implement the annual program spending requirements as written in the Small Business Act, in part because SBA's SBIR and STTR policy directives inaccurately summarize the law.

Specifically, the policy directives say that the relevant statutory provision requires participating agencies to "reserve" the minimum percentages for awards to small businesses each year.

Some of the participating agencies receive multiyear appropriations and choose to carry over some funding from one fiscal year to the next, in part because of this misinterpretation of the spending requirements. For example, if the amount of funding remaining in the program account is less than the amount of a phase I or phase II award, program managers said they carry over the funding to the next year, rather than make a smaller award to a recipient. An agency can carry over funding from one year to the next and comply with spending requirements if the agency spends the minimum required amount during the fiscal year, regardless of the year the funding was appropriated.

Carrying over funds to the next fiscal year without spending the minimum required amount contributed to two agencies—DOD and USDA—not complying with spending requirements in fiscal year 2012. Program managers from DOD and USDA said their agencies did not meet spending requirements using our method of calculating compliance, but DOD and USDA officials said they will spend all of the funding that was budgeted for the programs before the funding expires, which they believe meets the intent of the law. Program managers from both agencies explained that they receive multiyear appropriations for research and development funding, which would generally allow them to carry over funds from one year to the next. However, if an agency carries over funding from one year to the next without spending the minimum required amount on the program in each fiscal year, then the agency is not in compliance with the spending requirements. According to SBA officials, SBA agrees with our methodology for calculating compliance and interprets the Small Business Act as requiring agencies to spend, rather than reserve, the required percentages each year.

Carrying over funds from one year to the next without spending the minimum required amount in fiscal year 2012 also contributed to at least one agency—DOD—not complying with spending requirements in prior years. To demonstrate the effect that carrying over funds can have, we made the following calculations for DOD for fiscal years 2006 to 2011 using two different methodologies for calculating program spending.[16] First, we determined the amount spent in a specific year, regardless of the year the funding was appropriated, from DOD's total program obligations data submitted to SBA; this is the method that we used to calculate compliance with spending requirements for this report and the method SBA uses. Second, we calculated the amount of the funding appropriated in a fiscal year that was spent in that year or the next year; this is how DOD believes it should determine its compliance. We found that DOD complied with SBIR spending requirements in 2 of the 6 years from fiscal years 2006 through 2011 and with STTR spending requirements in 4 of the 6 years using our method. DOD's method is incorrect; however, if it were correct, it would yield a compliance determination in each of the 6 years for SBIR and in 5 years for STTR (see app. III for additional detail on our analysis).

Spending Requirements Increased Partway through the Year

Increases to the spending requirements partway through fiscal year 2012 contributed to two agencies not complying with spending requirements in fiscal year 2012, according to program managers. The 2011 reauthorization of the SBIR and STTR programs increased the minimum percentages to be spent for the programs for fiscal year 2012, but the law was enacted on December 31, 2011—a full quarter into fiscal year 2012. These increases were the first increases in many years for the programs: the SBIR levels had remained at a minimum of 2.5 percent of the extramural R&D budget from 1997 through 2011, and STTR levels had remained at a minimum of 0.3 percent from 2004 through 2011. Appropriations for fiscal year 2012 were enacted in November and December 2011, so some agencies had already planned their spending for the programs and had started to make awards to meet the prior percentages of 2.5 percent of their extramural R&D budgets for SBIR awards and 0.3 percent for STTR awards. For some agencies, this increase totaled millions of dollars. For example, NASA's SBIR spending requirement increased by about $5.5 million, and its STTR spending requirement increased by about $2.7 million over its estimates at the beginning of the fiscal year due to the higher percentages included in the reauthorization. If the spending requirements had remained at fiscal year 2011 levels, one of the three agencies that did not meet SBIR spending requirements for the year would have met the requirements, and the remaining two agencies would have been much closer to meeting the SBIR spending requirements, and all five agencies would have met the STTR requirements.

The increased spending requirements also affected agencies that complied with spending requirements for the year. For example, program managers at one agency that met SBIR spending requirements told us it was difficult to spend SBIR and STTR funds at the increased levels in the remaining portion of the fiscal year because the agencies had already begun to make awards, and their spending plans for the programs included funding a certain number of projects at a certain level. Program officials at one agency said that, as a result of the increased spending requirements, they increased the amounts of their awards for fiscal year 2013. The officials anticipate that the increased award amounts will help them meet their minimum spending requirements in future years. Program managers from two agencies that met SBIR spending requirements also told us it was difficult to obtain additional funding within their agency to meet the increased spending requirements for the programs because the reauthorization act and resulting increased spending requirements became effective after appropriations were enacted for the year.

Agencies Did Not Submit Complete Information on Program Spending

Program managers at two agencies said that they did not include technical assistance funds as part of their total spending data for the SBIR program, which contributed to the appearance that agencies were more out of compliance than they were for fiscal year 2012. Program managers from three other agencies also told us they could not submit these data. In certain circumstances, amounts spent for discretionary technical assistance count as part of the agency's funding for SBIR or STTR.[17] However, SBA's policy directives for the programs call for agencies to submit data on the amounts spent for awards, but not the amount spent on technical assistance. According to SBA officials, they did not request information on the amounts that agencies spent for technical assistance because SBA does not report the information to Congress.

Program officials at seven agencies—DOD, DOE, DHS, EPA, HHS, NSF, and USDA—said that they provided between about $125,000 and $3.5 million for discretionary technical assistance to certain SBIR or STTR award recipients in fiscal year 2012. Only two of the agencies—DOD and NSF—included this assistance in their total spending data.[18] Several program managers indicated that they did not submit data to SBA on the amounts spent for discretionary technical assistance because SBA did not provide written guidance on whether or how such spending should be included in their data. As a result, agencies did not report the total amounts that they spent on the programs and, therefore, Congress does not have access to complete and accurate information about the amounts spent on the SBIR and STTR programs. For the two agencies that provided technical assistance to award recipients but did not comply with spending requirements, including the amount spent on technical assistance in their total spending figures would have brought the agencies closer to meeting their spending requirements, but it would not have brought the agencies into compliance for the year.

Extramural R&D Was Higher Than Estimated at the Beginning of the Fiscal Year

Program officials at one agency said that changes to the amount of the agency's extramural R&D budget between the estimates early in the fiscal year and the actual amount obligated, which is determined at the end of the year, contributed to the agency not complying with the spending requirements. The Small Business Act requires agencies to calculate their extramural R&D budget by subtracting amounts obligated for intramural R&D from total obligations for R&D. However, the total amount obligated for extramural R&D is not known until the end of the fiscal year. For fiscal year 2012, NASA estimated its extramural R&D budget at the beginning of the year as about $5.5 billion and developed a spending plan for the SBIR and STTR programs based on this budget. However, according to program officials, NASA's extramural R&D budget increased to about $5.7 billion based on the total amount obligated for extramural R&D during the fiscal year. This increase in the extramural R&D budget throughout the year led to an unexpected increase in NASA's minimum spending requirements of about $6.3 million. The officials said that they had no way of knowing what the final extramural R&D budget number would be until after the end of the fiscal year, and by then it was too late to adjust their spending plan for the SBIR or STTR programs. As a result, the increase contributed to NASA's noncompliance with spending requirements in fiscal year 2012.

AGENCIES DID NOT FULLY COMPLY WITH CERTAIN REPORTING REQUIREMENTS AND SUBMITTED INCORRECT DATA

Participating agencies and SBA did not fully comply with certain reporting requirements for the SBIR and STTR programs. Specifically, participating agencies did not comply with requirements related to reporting their methodologies for calculating their extramural R&D budgets, and two agencies submitted incorrect data to SBA on the size of their extramural R&D budgets for the STTR program. Additionally, SBA has not yet submitted its report on the programs to Congress for fiscal year 2012.

Agencies Did Not Fully Comply with Methodology Reporting Requirements

Each of the agencies submitted its required report on the methodology for calculating the extramural R&D budget to SBA later than required by the Small Business Act and at least 6 of the agencies did not itemize and explain all exclusions from their calculations of their extramural R&D budgets, as called for in the policy directives.[19] Agencies are required by the Small Business Act to submit their methodology reports to SBA within 4 months of enactment of their annual appropriations. However, for fiscal year 2012, the 11 agencies included their methodology reports as part of their data submissions to SBA. Agencies generally submitted their data to SBA from March to July 2013—more than a year after agency appropriations for fiscal year 2012 were enacted on November 18, 2011, and December 23, 2011—because SBA allowed them to submit the reports later.[20] SBA officials said that, as in previous years, they did not need the methodology reports before the program data were due because the officials use the methodology reports to prepare SBA's report to Congress, which cannot be completed without the program data.

As we found in our first annual report on these issues, as a result of not having the methodology reports earlier in the year, SBA does not have an opportunity to promptly analyze these methodologies and provide the agencies with timely feedback to assist agencies in accurately calculating their spending requirements.[21] Without such review and feedback, agencies may be calculating their extramural R&D budgets incorrectly, which could lead to agencies spending less than the required amounts on the programs. We recommended in that report that SBA provide timely annual feedback to each agency following submission of its methodology report on whether its method for calculating the extramural R&D budget complies with program requirements. SBA officials said as of May 2014 that they plan to provide feedback to agencies on their methodology reports for fiscal year 2013 shortly after they receive the reports from the agencies, which are expected in June 2014.

Additionally, at least six agencies did not itemize and explain exclusions from their calculations of their extramural R&D budgets. Agencies are required to exclude certain subunits or programs in calculating the extramural R&D budget. For example, agencies with subunits in the intelligence community are to exclude those subunits from their extramural R&D budget.[22] SBA's policy directives call for agencies to itemize their exclusions and their reasons for such exclusions in their reports to SBA. Six of the 11 agencies' methodology reports—DOD, DOE, EPA, NASA, NSF, and DOT—noted that the agencies excluded some programs from their calculations, but these agencies either did not itemize the specific programs that they excluded, did not specify the reasons why they excluded the programs, or both. In the case of DOD, the agency's report stated that some of its programs were exempted by the section of the Small Business Act described above, which exempts programs in the intelligence community. However, DOD's report does not itemize the specific programs or subunits that were excluded, as required by the policy directives. Also, DOE, EPA, NASA, and NSF provided general categories of exclusions but not particular itemized programs. DOT excluded the research expenditures of the Federal Aviation Administration and the Federal Highway Administration from its extramural R&D budget without explanation in the report about why the programs were excluded.

As we found in our first annual report on these issues, agencies submitted different levels of detail on their methodologies, such as the programs excluded from the extramural budget and the reasons for the exclusions, and SBA did not raise questions about details of the methodologies.[23] SBA officials told us they would like agencies to include more information about exclusions in their methodology reports. In our first annual report, we concluded that, without guidance from SBA, participating agencies are likely to continue to provide SBA with broad, incomplete, or inconsistent information on their methodologies for calculating their extramural R&D budgets. At that time, we recommended that SBA provide additional guidance on the format agencies are to include in their methodology reports, among other issues. As of May 2014, SBA said that a working group of program managers had developed a template for agencies to use to submit their methodology reports. Agencies will be encouraged to use the template for their fiscal year 2013 submission and required to use it beginning in fiscal year 2014.

Two Agencies Submitted Incorrect Data for the STTR Program

Two agencies—DOD and HHS—submitted incorrect figures to SBA for the size of their extramural R&D budgets for the STTR program for fiscal year 2012.[24] The Small Business Act requires federal agencies to spend a certain amount of their extramural R&D budgets on the SBIR and STTR programs if the agency meets certain thresholds for participation. The threshold for participating in the programs is based on the entire agency's extramural R&D budget and not that of the specific subunits within the agency. Additionally, the extramural R&D budget figures for the SBIR and STTR programs should be consistent because the budget must be calculated the same way for both programs under the act. However, DOD submitted an extramural R&D budget for SBIR that was about $2.7 billion more than its budget for STTR, and HHS submitted an extramural R&D budget for SBIR that was about $337 million more than its budget for STTR. This occurred because the two agencies incorrectly excluded the budgets of subunits within the agencies that spent less than $1 billion per year on extramural R&D from the budget calculation for the STTR program. For example, HHS calculated its extramural R&D budget data for STTR using the amount of the extramural R&D budget of the National Institutes of Health and excluded the extramural R&D funding from the other subunits that participate in the SBIR program—the Centers for Disease Control and Prevention, the Food and Drug Administration, and the Administration for Children and Families—from its calculation.

As a result of the incorrect data, DOD and HHS underestimated their STTR spending requirements by about $9.3 million and $1.2 million, respectively, for fiscal year 2012. Using the correct extramural R&D budget figures did not change our assessment of DOD's or HHS's compliance with spending requirements for fiscal year 2012 because DOD did not meet its spending requirement using the lower extramural R&D budget figure and HHS's spending for the STTR program was high enough that it met its spending requirements even using the higher extramural R&D budget amount from its SBIR report. However, since spending requirements are calculated as a percentage of the extramural R&D budget, using an incorrect budget figure could lead to an agency not spending the required amount for the programs in the future. For example, the National Institutes of Health is the only subunit that participated in HHS's STTR program in fiscal year 2012. If the National Institutes of Health had not exceeded the minimum spending requirements for HHS in fiscal year 2012, HHS would not have complied with its agencywide spending requirements.

DOD officials said they were aware of the problem because we identified it in our first annual report, but it was too late for them to correct it for fiscal year 2012. The officials said that they have corrected the problem and plan to use the same extramural R&D budget numbers for the SBIR and STTR programs starting with the data submitted for fiscal year 2013. HHS officials told us they did not correct the budget numbers for fiscal year 2012 because it was too late to do so. However, they also said that they have not corrected this problem for fiscal years 2013 or 2014 because HHS has not received guidance from SBA to do so. SBA officials said they have discussed this issue with HHS. According to SBA's report to Congress for fiscal years 2009 through 2011, SBA and HHS were discussing the appropriate extramural budget figure.

SBA Has Not Submitted Its Annual Report for Fiscal Year 2012

SBA has not issued its report to Congress on the programs for fiscal year 2012. The Small Business Act requires SBA to report to certain congressional committees on the SBIR and STTR programs—including an analysis of the agencies' methodology reports—not less than annually, but the act does not specify a date that the report is due. SBA officials said that they had begun drafting the report for fiscal year 2012 but, early in 2014 two agencies—DOD and EPA—notified SBA that they needed to submit updated data, so SBA postponed its work on the report. According to SBA officials, SBA's report to Congress depends on the timeliness and quality of the participating agencies' submissions to SBA, and they delayed the submission of the last report and the fiscal year 2012 report to provide Congress with more accurate information on the SBIR and STTR programs. SBA officials told us they planned to submit their report for fiscal year 2012 by the end of fiscal year 2014—2 years after the subject fiscal year.

In our first annual report on these issues, we found that SBA had issued reports on 3 of the 6 years covered by our review—2006 through 2008— and those reports contained limited analyses of the agencies' methodologies, and some of the analyses were inaccurate.[25] SBA issued its report to Congress for fiscal years 2009 through 2011 in December 2013, making the data available to Congress about 2 to 4 years after the end of the subject year. In that report, we concluded that, without more rigorous oversight by SBA, and more timely and detailed reporting on the part of both SBA and participating agencies, it will be difficult for SBA to ensure that intended benefits of these programs are being attained and that Congress receives critical information to oversee these programs. Consequently, we recommended that SBA provide Congress with a timely annual report that includes a comprehensive analysis of the methodology each agency used for calculating the SBIR and STTR spending requirements. SBA agreed with our recommendation and stated in its comments on the report that it planned to implement the recommendation. SBA's policy directives for the programs call for agencies to submit their program data to SBA by March 15 each year.[26] However, SBA officials said that they have no enforcement mechanism to ensure that agencies submit their final data by that date. SBA introduced a scorecard in its report to Congress for fiscal years 2009 through 2011 to document agencies' timeliness in submitting data for fiscal year 2011. SBA officials expect that publishing the information in the report to Congress will encourage the agencies to submit their data on time, which, according to the officials, should improve SBA's timeliness in reporting to Congress.

CHANGING THE CALCULATION METHODOLOGY COULD INCREASE SPENDING REQUIREMENTS AND PARTICIPATION, WITH POTENTIAL BENEFITS AND DRAWBACKS

Potential effects of basing each participating agency's spending requirement on its total R&D budget instead of its extramural R&D budget include an increase in the amount of the spending requirement—for some agencies more than others—and, if the thresholds for participation in the programs did not change, an increase in the number of agencies required to participate. Agency officials identified some benefits and drawbacks of changing the methodology for calculating the spending requirements for the SBIR and STTR programs.

Changing the Calculation Methodology Could Increase Spending Requirements and Participation in the Programs

If the methodology for calculating the spending requirements had been changed so that the same percentages and thresholds for participating in the programs as under current law were applied to an agency's total R&D budget rather than to its extramural R&D budget in fiscal year 2012, all but one agency would have been required to spend more than under current law, according to our analysis of budget data and data submitted to SBA. [27] The change would have had this effect because an agency's extramural R&D budget is a part of and, therefore, smaller than its total R&D budget, and the same percentages would have been applied to a larger amount of funding. Figure 3 shows a comparison of the agencies' spending requirements for the

SBIR and STTR programs in fiscal year 2012 under the current law, based on an agency's extramural R&D budget, and this alternative methodology. These figures are generally consistent with the findings in our first annual report on these issues.

Agency (Dollars in millions)	Agency's budget: Total research and development budget[a]	Agency's budget: Extramural research or research and development budget[b]	Small Business Innovation Research Program: Spending requirement under current law[c]	Small Business Innovation Research Program: Spending requirement under alternative scenario[d]	Small Business Innovation Research Program: Percentage difference between current law and alternative scenario	Small Business Technology Transfer Program: Spending requirement under current law[e]	Small Business Technology Transfer Program: Spending requirement under alternative scenario[f]	Small Business Technology Transfer Program: Percentage difference between current law and alternative scenario
DOD	$72,811	$38,816	$1,009	$1,893	88%	$136	$255	88%
HHS	31,209	24,763	644	811	26	87	109	26
NASA	11,175	5,728	149	291	95	20	39	95
DOE	9,942	6,143	160	258	62	22	35	62
NSF	5,101	5,111	133	133	0g	18	18	0g
USDA	2,242	746	19	58	200	h (shaded)	8 (shaded)	h (shaded)
Veterans Affairs	1,160	h	h (shaded)	30 (shaded)	h (shaded)	h (shaded)	4 (shaded)	h (shaded)
Commerce	1,023	175	5	27	485	h (shaded)	4 (shaded)	h (shaded)
DOT	896	330	9	23	172	i	i	i
Interior	817	h	h (shaded)	21 (shaded)	h (shaded)	i	i	i
EPA	563	174	5	15	223	i	i	i
Education	397	327	8	10	22	i	i	i
DHS	384	322	8	10	19	i	i	i
Total	**$137,720**	**$82,634**	**$2,148**	**$3,581**	**67%**	**$282**	**$471**	**67%**

(Shaded:) Agency did not participate in the program in fiscal year 2012, but it would be required to do so if participation was based on the total research and development budget, and the thresholds for participating in the programs remained the same.

Source: GAO analysis of agency data. | GAO-14-431.

Note: Calculations may be affected by rounding.

[a]Each agency's total research and development budget authority is from the *Analytical Perspectives* volume of the President's budget. The category of "Facilities and Equipment" is not included in the total shown in this table.

[b]Each agency's extramural research or research and development (R&D) budget is the amount the agency submitted to SBA.

[c]Each agency's spending requirement under the current law for the Small Business Innovation Research (SBIR) program is calculated as 2.6 percent of its extramural R&D budget.

[d]Each agency's SBIR spending requirement under the alternative scenario is calculated as 2.6 percent of the total research and development budget authority and assumes no reductions for excluded programs.

[e]Each agency's spending requirement under the current law for the Small Business Technology Transfer (STTR) program is calculated as 0.35 percent of its extramural R&D budget.

[f]Each agency's STTR spending requirement under the alternative scenario is calculated as 0.35 percent of the total research and development budget authority and assumes no reductions for excluded programs.

[g]NSF's total research and development budget authority is less than its extramural R&D budget due to differences in the sources of data, according to NSF officials. NSF officials said that the amount reported in the *Analytical Perspectives* is based only on the actual budget authority for that particular year, while the amount submitted to SBA is based on total obligations resulting from budget authority provided in previous years.

[h]Agency did not participate in the program in fiscal year 2012.

[i]The agency's total research and development budget authority did not meet the threshold for participating in the STTR program for this alternative scenario.

Figure 3. Comparison of Fiscal Year 2012 Spending Requirements under the Current Law and an Alternative Scenario Based on Agencies' Total Research and Development Budgets.

As shown in figure 3, some agencies' spending requirements would increase more than others' under the alternative scenario. This variation is due to differences in the relative proportions of the extramural and intramural R&D budgets among agencies. Under the alternative scenario, agencies with a larger proportion of their total R&D budget for extramural R&D would experience a smaller increase in their spending requirements relative to the other agencies, as discussed in the following examples:

- DHS used about 84 percent of its total R&D budget for extramural R&D in fiscal year 2012 and was required to spend about $8 million on its SBIR program in that year. The agency would have a relatively small increase in its SBIR spending if the spending requirement were based on its total R&D budget instead of its extramural R&D budget: DHS would have been required to spend about $2 million more in fiscal year 2012 if the calculation methodology changed.
- The Department of Commerce used a relatively small percentage of its total R&D budget—about 17 percent—for extramural R&D in fiscal year 2012, and its spending requirement for SBIR would have increased more than five times, from about $5 million to $27 million in fiscal year 2012 if the calculation methodology had changed. While Commerce did not participate in the STTR program in fiscal year 2012, it would have been required to participate in the STTR program if the calculation methodology had changed and the thresholds for participating in the program had remained the same. Its spending requirement for the STTR program would have been $4 million rather than $0.

Consequently, if the funding percentage required in law in fiscal year 2012 applied to the total R&D budget instead of the extramural budget, Commerce's SBIR and STTR spending would have increased by a total of about $26 million.

Changing the calculation methodology to the total R&D budget instead of the extramural R&D budget would also increase the number of agencies that would be required to participate in the SBIR and STTR programs, assuming the dollar thresholds for participating in the programs remained the same. Two additional agencies—the Departments of Veterans Affairs and the Interior—would have been required to participate in the SBIR program in fiscal year 2012 if the dollar thresholds for participating in the programs applied to the total R&D budgets.[28] Adding these two agencies to the other agencies that participated in the SBIR program in fiscal year 2012 would have increased total federal SBIR spending requirements by about $51 million, in addition to the increased spending requirements for the other agencies. For STTR, three additional agencies—Commerce, USDA, and the Department of Veterans Affairs—would have been required to participate in the STTR program for fiscal year 2012 if the dollar thresholds for participating in the programs were applied to the total R&D budgets.[29]

Adding these three agencies to the other agencies that participated in the STTR program in fiscal year 2012 would have increased total federal STTR spending requirements by about $16 million, in addition to the increased spending requirements for the other agencies.

We recognize that other scenarios could have different effects on the spending requirements. For example, basing the SBIR and STTR spending requirement on an agency's total R&D budget and applying a lower percentage than under current law could result in reduced spending requirements for some agencies. Specifically, if the percentage applied to the total R&D budget had been 1.5 percent instead of 2.6 percent, and the thresholds for participating in the programs had remained the same, the spending requirement for the total SBIR programs would have slightly decreased in fiscal year 2012.[30]

Using this lower percentage, the spending requirements for some agencies, such as Commerce and DOD, would have increased, while the spending requirements for other agencies, such as DHS and NSF, would have decreased.

Program Managers Identified Benefits and Drawbacks of Changing the Calculation Methodology

Program managers identified benefits and drawbacks that basing the methodology for calculating the spending requirements on the total R&D budget could have for their agencies' programs. For example, several program managers indicated that more funding for the programs would allow their agencies to fund more projects, which was viewed as a benefit because the agencies tend to receive more proposals than they can fund in a year under current budgets. Additionally, several program managers told us that basing the SBIR and STTR spending requirements on the total R&D budget could improve transparency and reduce complexity in determining the SBIR and STTR spending requirements because the agencies would no longer need to identify the extramural portion of their R&D budgets, which the agencies have to determine internally. Similarly, program managers at some agencies said that the calculation to determine the agency's extramural R&D budget can be complex or difficult and that avoiding it would save time and effort, which would be a benefit.

Program managers also identified drawbacks to changing the methodology for calculating the spending requirements, with most program managers indicating that the potential drawbacks of such a change could outweigh the benefits. For example, most program managers said that increasing the amount of funding for the SBIR and STTR programs could be a drawback because funding used to support the agency's other R&D efforts would have to be spent on the SBIR and STTR programs. In addition, most program managers said that changing the methodology for calculating spending requirements for these programs could affect agency operations, depending on how the change is implemented within the agency. For example, several program managers said that basing spending requirements on the total R&D budget could increase SBIR and STTR budgets, which may lead to a decision within the agency to reduce intramural R&D programs and, in turn, lead to reductions in staffing of these programs. Most program managers also said that basing spending requirements on the total R&D budget could mean shifting money from other extramural R&D projects currently funded within the agency to SBIR and STTR.

Little Information Is Known About Total Administrative Costs

Little is known about the total amounts agencies spent administering the SBIR and STTR programs because agencies do not consistently collect such information. Agencies are not required to track or estimate all costs for administering the programs. In response to our requests to agencies for data on their fiscal year 2012 administrative costs, most agencies provided information on some categories of administrative costs and partial estimates of costs.[31] Agencies provided information on several types of administrative costs, such as staff, contractor, and travel costs for preparing and awarding contracts and grants and for monitoring them. Administrative cost estimates that the agencies provided for fiscal year 2012 ranged from about $200,000 to about $8 million. As with the data on fiscal year 2011 for our first annual report, these data were wide-ranging, incomplete, and unverifiable.

The 2011 reauthorization of the programs created a pilot program beginning in fiscal year 2013 that allows agencies to spend up to 3 percent of SBIR funds for administrative costs of the programs beginning in fiscal year 2013. The SBIR policy directive requires agencies to submit information to SBA on the funding used for administrative costs through the pilot program. As such, we anticipate that agencies that participate in the pilot program will have additional information on administrative costs for fiscal year 2013, which we will examine in our next annual review of the programs. As of May 2014, SBA officials expected that agencies would submit their data for fiscal year 2013 in June 2014. According to SBA's policy directives, funding under the pilot program must not replace current agency administrative funding for SBIR. Consequently, even with the pilot program, agencies will likely not identify or track all administrative costs for SBIR.

Conclusion

Federal agencies have awarded billions of dollars to small businesses under the SBIR and STTR programs to develop and commercialize innovative technologies. In our first annual report on these issues, we identified some areas where SBA could take actions to ensure that agencies comply with spending and reporting requirements. Our report was issued in September 2013, after agencies' fiscal year 2012 data and reports were due to SBA, and, as a result, some of the issues we identified in the data and reports for fiscal years 2006 through 2011 remained in fiscal year 2012. For example, in our prior report, we found that the agencies submitted different levels of detail in their methodology reports, such as the programs excluded from the extramural R&D budget and the reasons for the exclusions. In that report, we recommended that the SBA Administrator provide the format for agencies to include in their methodology reports and to provide timely annual feedback to agencies on whether their methodology calculations for extramural R&D budgets comply with program requirements. We found similar issues with agencies' reports for fiscal year 2012 and continue to believe that this oversight is needed and that the recommendation is valid and should be fully implemented. Additionally, as part of its oversight role, SBA is required to provide Congress with annual reports on the SBIR and STTR programs. However, SBA has faced challenges in submitting its report in a timely manner—particularly with agencies revising their data submissions well after the fiscal year is over—leading to reports that are submitted between 2 and 4 years after the subject year. In our September 2013 report, we recommended that SBA provide Congress with a timely annual report that includes a comprehensive analysis of the methodology each agency used for calculating the SBIR and STTR spending requirements. We continue to believe that providing a timely report to Congress is important and that the recommendation continues to have merit and should be fully implemented. Moreover, such a report could demonstrate SBA's oversight to help ensure that the programs are meeting their intended goals.

In addition to those issues that could be addressed by implementing our prior recommendations, we identified four additional issues that—if left unaddressed—could affect compliance with spending and reporting requirements in the future. First, some agencies did not comply with the requirement to spend program funds annually because SBA's policy directives inaccurately summarize the requirement. Without correct information in the policy directives, such misinterpretations could lead to agencies spending less than required. Second, agencies did not report the total amounts spent on the programs because SBA has not provided guidance on whether or how to submit information on spending for reasons other than awards, such as technical assistance. As a result, Congress does not have access to complete and accurate information on the amounts spent on the programs. Third, agencies have continued to submit their methodology reports to SBA significantly later than required by law. In fiscal year 2012, most agencies submitted their methodology reports more than a year after the deadline established in the Small Business Act. Without holding agencies to the deadlines in the law, SBA did not have an opportunity to perform timely oversight by analyzing these methodologies and providing the agencies with feedback, if needed, to correct problems with the calculations and help to ensure that agencies meet mandated spending requirements. Fourth, two agencies have calculated their extramural R&D budgets for the STTR program incorrectly. DOD officials told us that they will use the same extramural R&D budget figure for both programs moving forward, but HHS officials do not plan to do so. Using the wrong figure could lead to spending less than the required amount for the programs in the future.

RECOMMENDATIONS FOR EXECUTIVE ACTION

To improve compliance with the Small Business Act and enhance SBA's ability to provide oversight of the programs, we recommend that the SBA Administrator take the following three actions:

- Revise the language in the SBIR and STTR policy directives to accurately summarize the statutory provisions that describe the program spending requirements.
- Provide written guidance to agencies clarifying whether and how agencies should submit data to SBA on spending outside of awards that is allowed under the programs, such as discretionary technical assistance.
- Request that the agencies submit their methodology reports within 4 months of the enactment of appropriations, as required by the Small Business Act and the program policy directives.

To help ensure that the agency continues to spend the required amount for the STTR program, we recommend that the Secretary of Health and Human Services include all of the agency's extramural R&D budget in the calculation of STTR spending requirements and in the data submitted to SBA.

AGENCY COMMENTS AND OUR EVALUATION

We provided a draft of this report to the Secretaries of Health and Human Services, Agriculture, Commerce, Defense, Education, Energy, Homeland Security, and Transportation; the Administrators of the Small Business Administration, the Environmental Protection Agency, and the National Aeronautics and Space Administration; and the Director of the National Science Foundation for review and comment. In response, six of the agencies Commerce, Education, DOE, DHS, DOT, and EPA— stated by email that they had no technical or written comments. Two agencies— NASA and NSF—provided technical comments by e-mail but did not provide formal written comments. We incorporated these technical comments, as appropriate. The remaining four agencies—SBA, HHS, USDA, and DOD—provided written comments, which are discussed below.

In written comments, SBA generally agreed with the issues and recommendations included in our report. The letter states that SBA does not believe that the policy directive inaccurately summarizes the law and that the use of the term “reserve” refers to the allocation of the extramural budget. SBA officials plan to start using the term “extramural obligations” instead of “extramural budget” to address confusion with participating agencies, but they will review all language in the policy directives to ensure that the funding requirements are clear. SBA’s proposed change could help address some confusion among participating agencies, but we continue to believe that SBA should implement our first recommendation to revise the language in the SBIR and STTR policy directives to accurately summarize the statutory provisions that describe the program spending requirements. Specifically, the policy directives currently say that the relevant statutory provision requires participating agencies to “reserve” a minimum percentage of their extramural budget for awards to small businesses each year, whereas the law requires participating agencies to “expend” the minimum percentages. As reflected in USDA’s agency comments (see app. VI), the use of the term “reserve” is a specific point of confusion for at least one of the participating agencies.

In written comments, HHS concurred with our recommendation regarding the inclusion of all of the agency’s extramural R&D budget in the calculation of the STTR spending requirements and in the data submitted to SBA. HHS plans to ensure that the agency’s extramural R&D budget is calculated the same way for the SBIR and STTR programs beginning in fiscal year 2015, as funding for the programs in fiscal year 2014 is already under way.

In written comments, USDA's Director of the National Institute of Food and Agriculture, stated that USDA generally agrees with our report, but the agency does not agree that it did not comply with the spending requirements for the SBIR program. USDA's written comments state that USDA has not always obligated the minimum percentage of its extramural R&D budget for the SBIR program in a single fiscal year because some of the funds used for the program are available without fiscal year limitations, referred to as "no-year money." USDA's written comments explain that USDA did not obligate the required amounts due to the language in the program policy directive, which directs federal agencies to "reserve" the minimum percentage of their extramural R&D budgets for awards to small business concerns. This is consistent with what officials told as during our review. We recognize that agencies with multiyear appropriations are generally allowed to carry funds over from one year to the next and that there are benefits to doing so, but we continue to believe that the Small Business Act requires participating agencies to spend at least the minimum percentage of their extramural R&D budgets on the program each year, regardless of the year the dollars spent were appropriated. In other words, an agency can use a mix of funds from different fiscal years, consistent with the limitations specific to its appropriations, to meet the spending requirement in any given fiscal year, but it must still meet the spending requirement in each fiscal year. Conversely, as we state in the report, if an agency carries over funding from one year to the next without spending, or at least obligating, the minimum required amount on the program in each fiscal year, then the agency is not in compliance with spending requirements. As noted above, we continue to believe that a change to the program policy directives is needed to address this misinterpretation of the spending requirements.

In written comments, the DOD Administrator for the SBIR and STTR programs stated that DOD concurs with our finding that DOD does not expend all SBIR and STTR funds appropriated to the department in the same fiscal year that they are received, but DOD believes that the agency is in compliance with the law in that it obligates the minimum percentage of SBIR and STTR funds each year in accordance with the obligation authority conferred by Congress and longstanding DOD procedures. We neither state in the report, nor do we believe, that agencies are required to spend all of the SBIR or STTR funds appropriated in the same year that they are received. As we state in the report, we used the total obligations data that agencies submitted to SBA to calculate compliance with spending requirements. These data represent the amount obligated in a specific year, regardless of the year the funding was appropriated. Additionally, we agree that an agency is in compliance with program spending requirements if it obligates the minimum percentage of program funds each year. However, using the spending data that DOD submitted to SBA for fiscal year 2012, DOD did not comply with the spending requirements for the STTR program.

Frank Rusco
Director, Natural Resources and Environment

APPENDIX I: AGENCIES' COMPLIANCE WITH SPENDING REQUIREMENTS FOR THE SMALL BUSINESS INNOVATION RESEARCH PROGRAM FOR FISCAL YEAR 2012

Our analysis of the data that the agencies submitted to the Small Business Administration (SBA) indicate that the amounts that 8 of the 11 participating agencies spent for the Small Business Innovation Research (SBIR) program met or exceeded their fiscal year 2012 spending requirements, while spending for the remaining 3 agencies did not meet the requirements. See table 2 for data on agency compliance.

Table 2. Agency Compliance with Small Business Innovation Research (SBIR) Program Spending Requirements for Fiscal Year 2012

Dollars in millions					
Agency	**Extramural research or research and development (R&D) budget**	**Amount spent[a]**	**Calculated spending requirement[b]**	**Difference between amount spent and spending requirement**	**Percentage of extramural R&D budget spent forSBIR**
Department of Defense (DOD)	$38,816	$1,080.3	$ 1,009.2	$ 71.1	2.8%
Department of Health and Human Services (HHS)	24,763	648.6c	643.8	4.7	2.6%
Department of Energy (DOE)	6,143	169. 8c	159.7	10.1	2.8%

Table 2. (Continued)

Dollars in millions					
Agency	**Extramural research or research and development (R&D) budget**	**Amount spent**[a]	**Calculated spending requirement**[b]	**Difference between amount spent and spending requirement**	**Percentage of extramural R&D budget spent forSBIR**
National Aeronautics and Space Administration (NASA)	5,728	140.0	148.9	(9.0)	2.4%
National Science Foundation (NSF)	5,111	131.3	132. 9	(1.6)d	2.6%
U.S. Department of Agriculture (USDA)	746	16.9	19.4	(2.5)	2.3%
Department of Education (Education)	327	13.1	8.5	4.6	4.0%
Department of Homeland Security (DHS)	322	12.9	8.4	4.5	4.0%
Department of Transportation (DOT)	330	9.1	8.6	0.6	2.8%
Department of Commerce (Commerce)	175	4.5	4.5e	(0.0)d	2.6%
Environmental Protection Agency (EPA)	174	4.3c	4.5	(0.2)	2.5%
Total	**$82,634**	**$2,230.7**	**$2,148.5**	**$82.3**	**2.7%**

Source: GAO analysis of agency data reported to SBA.

Notes: Calculations may be affected by rounding.

Cells shaded gray indicate the agency did not meet its spending requirement in fiscal year 2012.

[a]The amount spent is the total obligations that the agency submitted to the Small Business Administration (SBA).

[b]We calculated the spending requirement as 2.6 percent of the agency's extramural R&D budget.

[c]Three agencies—DOE, EPA, and HHS—provided us revised spending data for fiscal year 2012 based on adding in their funding for discretionary technical assistance. We did not include the revised
data in our analysis because we used the data that agencies submitted to SBA and the agencies had not provided the updated data to SBA at the time of our review.
[d]If an agency's spending for the SBIR program as a percentage of its extramural R&D budget was within a rounding error of the required level, we considered that agency to be in compliance with the spending requirement. This was the case with NSF and Commerce. In the latter case, the amount spent was slightly below $4.5 million.
[e]Officials from one subunit within Commerce—the National Institute of Standards and Technology— said that they calculate the spending requirement for the program based on its actual spending on extramural R&D for the previous fiscal year, rather than the extramural R&D budget for the current fiscal year, but did not report this figure to SBA. We used the data that Commerce submitted to SBA to calculate its spending requirements.

APPENDIX II: AGENCIES' COMPLIANCE WITH SPENDING REQUIREMENTS FOR THE SMALL BUSINESS TECHNOLOGY TRANSFER PROGRAM FOR FISCAL YEAR 2012

Our analysis of the data that the agencies submitted to the Small Business Administration (SBA) indicate that two of the five agencies participating in the Small Business Technology Transfer (STTR) program met or exceeded their fiscal year 2012 spending requirements, while the remaining three agencies did not. See table 3 for data on agency compliance.

Table 3. Agency Compliance with Small Business Technology Transfer (STTR) Program Spending Requirements for Fiscal Year 2012

Dollars in millions					
Agency	**Extramural research or research and development (R&D) budget**	**Amount spent[a]**	**Calculated spending requirement[b]**	**Difference between amount spent and spending requirement**	**Percentage of extramural R&D budget spent forSTTR**
Department of Defense (DOD)	$38,816c	$118.8	$135.9	$(17.0)	0.31%

Table 3. (Continued)

Dollars in millions					
Agency	**Extramural research or research and development (R&D) budget**	**Amount spent[a]**	**Calculated spending requirement[b]**	**Difference between amount spent and spending requirement**	**Percentage of extramural R&D budget spent forSTTR**
Department of Health and Human Services (HHS)	24,763c	86.9	86.7	0.3	0.35%
Department of Energy (DOE)	6,143	23.5	21.5	2.0	0.38%
National Space and Aeronautics Administra-tion (NASA)	5,728	18.5	20.0	(1.5)	0.32%
National Science Foundation (NSF)	5,111	15.6	17.9	(2.3)	0.30%
Total	**$80,560**	**$ 263.3**	**$282.0**	**$(18.6)**	**0.32%**

Notes: Calculations may be affected by rounding.

Cells shaded gray indicate the agency did not meet its spending requirement in fiscal year 2012.

[a]The amount spent is the total obligations that the agency submitted to the Small Business Administration (SBA).

[b]We calculated the spending requirement as 0.35 percent of the agency's extramural R&D budget.

[c]We found that two agencies—DOD and HHS—submitted different figures to SBA for their extramural R&D budgets for the SBIR and STTR programs for fiscal year 2012. The difference in the extramural R&D budget figures would not have affected our assessment of DOD's or HHS's compliance with spending requirements for fiscal year 2012. Specifically, DOD did not meet its STTR spending requirements using either the higher or lower extramural R&D budget figures, and HHS's spending for the STTR program was high enough that it met its spending requirements even using the higher figure. Because we determined that the STTR extramural R&D budget figures were incorrect for DOD and HHS, we used the SBIR extramural R&D budget figures to determine DOD's and HHS's compliance with STTR spending requirements for this report.

APPENDIX III: AFFECT OF DIFFERENT METHODOLOGIES ON CALCULATION OF DEPARTMENT OF DEFENSE SPENDING FOR FISCAL YEARS 2006 THROUGH 2011

Differences in how agencies interpret the spending requirements for the Small Business Innovation Research (SBIR) and Small Business Technology Transfer (STTR) programs can lead to different results when calculating compliance with spending requirements. To demonstrate the effect that these differences can have, we made the following calculations for the Department of Defense (DOD). First, we used the methodology we used for this report, as well as our first annual report on these issues, in which we compared the agency's spending in each year to the spending requirements for that year. Second, we used the methodology that DOD uses to calculate its spending. DOD calculates its spending for the year as the funding that was appropriated in one year and spent in the same year and the following year. For example, for fiscal year 2006, DOD calculated its spending as the sum of the funding appropriated in fiscal year 2006 and spent in fiscal year 2006 and the funding appropriated in fiscal year 2006 and spent in fiscal year 2007. We used DOD for this comparison because DOD is the only agency that submits its spending data to the Small Business Administration (SBA) in a way that allows us to determine the year that funding was appropriated and spent.[1] We found that DOD complied with SBIR spending requirements in 2 of the 6 fiscal years from 2006 to 2011 and with STTR spending requirements in 4 of the 6 years using our method. DOD's method is incorrect, but if it were correct, it would yield a compliance determination in each of the 6 years for both the SBIR and STTR programs (see tables 4 and 5).

Table 4. Comparison of GAO and DOD Calculations for Small Business Innovation Research (SBIR) Program Spending for Fiscal Years 2006 through 2011

Dollars in millions						
Year	Extramural research or research and development (R&D) budget	Spending requirement[a]	GAO's method		DOD's method	DOD's method
			Dollars spent	Spending as a percentage of extramural R&D budget	Dollars spent	Spending as a percentage of extramural R&D budget
2006	$45,437	$1,136	$1,159	2.6%	$1,196	2.6%
2007	46,782	1,170	1,107	2.4%	1,218	2.6%

Table 4. (Continued)

Dollars in millions						
Year	Extramural research or research and development (R&D) budget	Spending requirement[a]	GAO's method		DOD's method	DOD's method
			Dollars spent	Spending as a percentage of extramural R&D budget	Dollars spent	Spending as a percentage of extramural R&D budget
2008	46,857	1,171	1,101	2.4%	1,230	2.6%
2009	48,884	1,222	1,150	2.4%	1,251	2.6%
2010	47,955	1,199	1,154	2.4%	1,241	2.6%
2011	$41,678	$1,042	$1,082	2.6%	$1,081	2.6%

Source: GAO analysis of agency data reported to the Small Business Administration (SBA).

Notes: Calculations may be affected by rounding.

Cells shaded gray indicate the agency did not meet its spending requirement.

These figures may differ from the figures presented in SBA's report to Congress for fiscal years 2009 through 2011 because DOD submitted updated to data to SBA after it provided the data to us and we needed the level of detail in the agency's data to SBA to perform the analysis.

[a]The spending requirement for fiscal years 2006 through 2011 was 2.5 percent of the extramural R&D budget.

Table 5. Comparison of GAO and DOD Calculations for Small Business Technology Transfer (STTR) Program Spending for Fiscal Years 2006 through 2011

Dollars in millions						
Year	Extramural research or research and development (R&D) budget	Spending requirement[a]	GAO's method		DOD's method	
			Dollars spent	Spending as a percentage of extramural R&D budget	Dollars spent	Spending as a percentage of extramural R&D budget
2006	$45,437	$130.9	$134.3	0.31%	$133.0	0.31%
2007	46,782	134.1	134.8	0.31%	137.7	0.31%
2008	46,857	133.8	116.9	0.26%	139.2	0.31%
2009	48,884	139.1	136.0	0.29%	137.2	0.30%
2010	47,955	135.2	149.6	0.33%	138.0	0.31%
2011	$41,678	$117.7	$120.9	0.31%	$122.6	0.31%

Source: GAO analysis of agency data reported to the Small Business Administration (SBA).

Notes: Calculations may be affected by rounding.

Cells shaded gray indicate the agency did not meet its spending requirement.

These figures may differ from the figures presented in SBA's report to Congress for fiscal years 2009 through 2011 because DOD submitted updated to data to SBA after it provided the data to us and we needed the level of detail in the agency's data to SBA to perform the analysis.

[a]The spending requirement for fiscal years 2006 through 2011 was 0.3 percent of the extramural R&D budget. Unlike the assessment in the body of the report, we used the extramural R&D budget figure DOD submitted for the STTR program because that was the amount the agency planned to spend for the year.

End Notes

[1] Agencies' R&D programs generally include funding for two types of R&D: intramural and extramural. Intramural R&D is conducted by employees of a federal agency in or through government-owned, government-operated facilities. Extramural R&D is generally conducted by nonfederal employees outside of federal facilities. Agencies are required to calculate their extramural R&D budget by subtracting amounts obligated for intramural R&D from total obligations for R&D.

[2] In this report, we refer to the amounts resulting from applying these mandated percentages to extramural R&D budgets as "spending requirements."

[3] National Defense Authorization Act for Fiscal Year 2012, Pub. L. No. 112-81, 125 Stat. 1298, § 5136.

[4] GAO, *Small Business Research Programs: Actions Needed to Improve Compliance with Spending and Reporting Requirements,* GAO-13-421 (Washington, D.C.: Sept. 9, 2013). In addition to our first annual report to address this mandate, we have issued several other reports on the SBIR and STTR programs dating back to the 1980s, including, most recently, *Small Business Innovation Research: Data Rights Protections,* GAO-14-116R (Washington, D.C.: Nov. 4, 2013) and *Small Business Research Programs: Agencies Are Implementing New Fraud, Waste, and Abuse Requirements,* GAO-13-70R (Washington, D.C.: Nov. 15, 2012).

[5] We used the agencies' total program obligations data to represent spending for the programs, in part because obligations data were readily available from each of the agencies for program purposes, and because obligations provided a reasonable measure of the spending for the programs in each year.

[6] If an agency's spending for the SBIR or STTR programs as a percentage of its extramural R&D budget was within a rounding error of the required level, we considered that agency to be in compliance with the spending requirement.

[7] Office of Management and Budget, *Fiscal Year 2014 Analytical Perspectives, Budget of the U.S. Government* (Washington, D.C.: Apr. 10, 2013). Because agencies did not include their total R&D budget authority in their data submitted to SBA, and we were seeking information across all government agencies, we relied on the amount of budget authority reported in the *Analytical Perspectives* volume for calculation of the spending requirements for the alternative scenarios.

[8] Research institutions include nonprofit colleges or universities and federally funded research and development centers.

[9] Pub. L. No. 112-81, Div. E.

[10] For examples of how one agency has used phase III awards, see GAO, *Small Business Innovation Research: DOD's Program Supports Weapon Systems, but Lacks Comprehensive Data on Technology Transition Outcomes,* GAO-14-96 (Washington, D.C.: Dec. 20, 2013).

[11] For the purposes of this report, we defined compliance as spending at least 2.6 percent of an agency's extramural R&D budget on the SBIR program in fiscal year 2012, as required by the Small Business Act. This method is consistent with SBA's approach for determining compliance with spending requirements in its annual report to Congress on the program. We used the agencies' obligations data to represent spending for the programs, in part because obligations data were readily available from each of the agencies for program purposes, and because obligations provided a reasonable measure of the spending for the programs in each year.

[12] See GAO-13-421. NSF was the agency that did not comply with SBIR spending requirements in fiscal year 2011. DHS, Education, and HHS complied with their SBIR spending requirements each fiscal year from 2006 to 2011. The agency data we reported differed from some of the data included in SBA's report to Congress for fiscal years 2009 to 2011, which SBA issued in December 2013. According to SBA officials, some agencies submitted updated data to SBA for those years after they provided the data to us. These changes did not affect the number of agencies that complied with spending requirements in fiscal year 2011 or in all 6 fiscal years from 2006 through 2011. For SBA's annual report to Congress, see *SBA: The Small Business Innovation Research (SBIR) and Small Business Technology Transfer (STTR) Program Annual Report Fiscal Years 2009-2011* (Washington, D.C.: December 2013).

[13] For the purposes of this report, we defined compliance as spending at least 0.35 percent of an agency's extramural R&D budget on the STTR program in fiscal year 2012, as required by the Small Business Act. This method is consistent with SBA's approach for determining compliance with spending requirements in its reports to Congress on the program. We used the agencies' obligations data to represent spending for the programs, in part because obligations data were readily available from each of the agencies for program purposes, and because obligations provided a reasonable measure of the spending for the programs in each year.

[14] See GAO-13-421. The agency data that we reported differed from some of the data included in SBA's report to Congress for fiscal years 2009 to 2011, which SBA issued in December 2013. According to SBA officials, some agencies submitted updated data to SBA for those years after they provided the data to us. These changes did not affect the agencies that complied with spending requirements in fiscal year 2011 or in all 6 fiscal years from 2006 through 2011. For SBA's report, see *SBA: SBIR and STTR Program Annual Report Fiscal Years 2009-2011*.

[15] We determined that an agency complied with spending requirements for the SBIR or STTR programs if the agency spent at least the minimum required percentage of the extramural R&D budget. We used the data that each agency submitted to SBA for total program obligations for fiscal year 2012, regardless of the year the funding was appropriated, to represent program spending for the year.

[16] We used DOD for this comparison because DOD is the only agency that submitted its spending data to SBA in a way that allowed us to determine the year that funding was appropriated and spent. We could not perform the analysis for DOD's fiscal year 2012 data because the amount of DOD's fiscal year 2012 appropriation that was obligated in fiscal year 2013 will

be included in DOD's fiscal year 2013 data submission to SBA, which had not yet been submitted at the time of our review. Agencies' data submissions are typically due to SBA in the March following the end of the fiscal year, but SBA officials said they extended the deadline for the fiscal year 2013 data submissions.

[17] Agencies are allowed to spend up to $5,000 of program funds per year for each phase I or phase II award through the SBIR or STTR programs for discretionary technical assistance. According to the program policy directives, the purpose of technical assistance is to assist the award recipients in making better technical decisions, solving technical problems that arise during the project, minimizing technical risks associated with the project, and commercializing the product of the research. This funding can be provided either to a vendor that provides the technical assistance to the award recipient or directly to the award recipient. The policy directives also state that funding for technical assistance will be in addition to the award, but counts as part of the agency's funding for SBIR or STTR, unless the agency funds the technical assistance using funds outside of those set aside for SBIR or STTR.

[18] Three agencies—DOE, EPA, and HHS—provided us revised spending data for fiscal year 2012 based on adding in their funding for discretionary technical assistance. We did not include the revised data in our analysis because we used the data that agencies submitted to SBA, and the agencies had not provided the updated total spending data to SBA at the time of our review.

[19] The remaining five agencies did not indicate in their methodology reports to SBA whether they had exclusions for some or all of their programs. Agencies are not explicitly required to do so. However, we cannot judge whether these agencies may have had exclusions that were not included in the report to SBA.

[20] See Consolidated and Further Continuing Appropriations Act, 2012, Pub. L. No. 112-55, 125 Stat. 552; Consolidated Appropriations Act, 2012, Pub. L. No. 112-74, 125 Stat. 786.

[21] GAO-13-421.

[22] 15 U.S.C. § 638 (e)(2) (2012).

[23] GAO-13-421.

[24] These agencies also submitted incorrect figures for their extramural R&D budgets for the STTR program in previous years, as we found in our September 2013 report. Because we determined that the STTR extramural R&D budget figures were incorrect, we used the SBIR extramural R&D budget figures, which were higher than the STTR figures, to determine DOD's and HHS's compliance with STTR spending requirements for this report.

[25] GAO-13-421.

[26] According to SBA officials, they extended the deadline for fiscal year 2013 data submissions. As of May 2014, SBA officials expected that agencies would submit their data for fiscal year 2013 in June 2014.

[27] NSF's total R&D budget authority in the *Analytical Perspectives* volume of the President's budget is less than its extramural R&D budget data submitted to SBA for fiscal year 2012 because the data are drawn from different sources, according to NSF officials. NSF officials said that the amount reported in the *Analytical Perspectives* is based only on the actual budget authority for that particular year, while the amount submitted to SBA is based on total obligations resulting from budget authority provided in previous years.

[28] Based on data in the President's budget, these two agencies had total R&D budget authority in excess of $100 million; federal agencies with a budget of $100 million or more for extramural R&D are required to establish and operate an SBIR program.

[29] These agencies reported total R&D budget authority in excess of $1 billion; federal agencies with budgets of $1 billion or more for extramural R&D are required to establish and operate an STTR program.

[30] We selected 1.5 percent as an alternative percentage because it is roughly equal to the average of the agencies' spending requirements for fiscal year 2012 as a percentage of the total R&D budgets.

[31] One additional agency—Commerce —provided administrative cost data for fiscal year 2012 that did not provide data for previous years. However officials at two agencies, HHS and DOD, said they could not provide administrative costs for fiscal year 2012. HHS officials said it would be very difficult to determine how much is spent on the administrative costs of the program since such funding for salaries, expenses, training, and travel, comes from accounts other than the SBIR and STTR programs. DOD officials said they do not have an estimate of administrative costs largely because thousands of employees throughout the department are involved with the SBIR and STTR programs and DOD does not have the ability to track the percentage of these employees' workload to support the two programs. HHS did not provide an estimate for fiscal year 2011 for our last report, but DOD provided an estimate of about $30 million for fiscal year 2011.

End Note for Appendix III

[1] We could not perform the analysis for DOD's fiscal year 2012 data because the amount of DOD's fiscal year 2012 appropriations that were obligated in fiscal year 2013 will be included in DOD's fiscal year 2013 data submission to SBA, which had not yet been submitted at the time of our review. Agencies' data submissions are typically due to SBA in the March following the end of the fiscal year, but SBA officials said they are extending the deadline for the fiscal year 2013 data submissions.

In: Small Business Innovation Research ... ISBN: 978-1-63482-532-0
Editor: Thomas H. Chavez

Chapter 3

SMALL BUSINESS INNOVATION RESEARCH: CHANGE IN PROGRAM ELIGIBILITY HAS HAD LITTLE IMPACT*

United States Government Accountability Office

ABBREVIATIONS

ARPA-E	Advanced Research Projects Agency-Energy
CDC	Centers for Disease Control and Prevention
DOE	Department of Energy
HHS	Department of Health and Human Services
NDAA	National Defense Authorization Act for Fiscal Year 2012
NIH	National Institutes of Health
R&D	research or research and development
SBA	Small Business Administration
SBIR	Small Business Innovation Research

* This is an edited, reformatted and augmented version of The United States Government Accountability Office publication, GAO-15-68, dated November 2014.

* * *

November 20, 2014

The Honorable Maria Cantwell
Chairwoman
The Honorable James E. Risch
Ranking Member
Committee on Small Business and Entrepreneurship
United States Senate
The Honorable Sam Graves Chairman
The Honorable Nydia M. Velázquez
Ranking Member
Committee on Small Business House of Representatives

Starting in 2013, implementation of the National Defense Authorization Act for Fiscal Year 2012 (NDAA) enabled agencies to allow small businesses that are majority owned by multiple venture capital operating companies, hedge funds, or private equity firms (majority-owned portfolio companies) to participate in the Small Business Innovation Research (SBIR) program. In 1982, Congress established the SBIR program to, among other things, stimulate technological innovation related to federal agency goals and missions, use small businesses to meet federal research or research and development (R&D) needs, and increase private sector commercialization of innovations derived from federal R&D.[1] In fiscal year 2013, federal agencies made more than 4,000 awards totaling more than $2.1 billion through the SBIR program. Over the program's first two decades, some majority-owned portfolio companies participated in SBIR, without any apparent adverse consequence for the program's operation and achievements, according to a study by the National Academy of Sciences.[2] However, in 2003, a Small Business Administration (SBA) administrative law judge issued a decision holding that such companies were ineligible to receive SBIR awards.[3] The decision raised concerns that some portfolio companies were being unfairly excluded from SBIR and that important research was not being funded. Others supported the decision, contending that small businesses without venture capital funding would not be able to compete against small businesses with such funding.

The NDAA reauthorized the SBIR program through fiscal year 2017 and included provisions that allow SBIR participating agencies to permit majority-owned portfolio companies to participate in their SBIR programs.[4] In

December 2012, SBA issued a final rule to implement the NDAA provisions and revise SBIR's eligibility criteria to include majority-owned portfolio companies, which became effective in January 2013.[5] The act allows but does not require agencies to permit majority-owned portfolio companies to participate in their SBIR programs. In order to allow such companies to participate, SBIR participating agencies must submit a written determination to SBA and Congress, explaining how making SBIR awards to such companies will, among other things, induce additional venture capital funding of small business innovations and substantially contribute to the agency's mission.[6] In addition, the NDAA mandates that GAO conduct a study of the impact of requirements related to participation in the SBIR program by majority-owned portfolio companies and issue a report every 3 years, with the first report due by December 31, 2014.[7] This report examines (1) the impact of allowing majority-owned portfolio companies to participate in agency SBIR programs and (2) the extent to which agencies have elected to expand their SBIR programs to include majority-owned portfolio companies. Our analysis of the impact focuses on determining which agencies opened their SBIR programs to majority-owned portfolio companies and how many of these companies applied for and received SBIR awards.

To address our objectives, we obtained and reviewed the written determinations submitted by the agencies to SBA and Congress—that is, those submitted by the Department of Energy (DOE) and National Institutes of Health (NIH). We also reviewed the NDAA and related congressional record and hearings on SBIR, SBA's proposed and final rules implementing the NDAA by revising the eligibility criteria for SBIR, including comment letters submitted in response to the proposed rule; and SBA's SBIR Policy Directive, annual reports, and other related material. We also reviewed and analyzed SBIR data on solicitations, applications, and awards from DOE and NIH during fiscal years 2013 and 2014, to identify awards made to majority-owned portfolio companies; data from the other nine SBIR participating agencies on SBIR awards they made during fiscal year 2013; and data on majority-owned portfolio companies and other small businesses registered with SBA. To assess the reliability of these data, we interviewed agency officials about the source of the data and data quality control procedures for the data and reviewed relevant documentation. We determined the data were sufficiently reliable for the purposes of this report. In addition, we reviewed studies on SBIR by the National Academy of Sciences, Congressional Research Service, and GAO. We interviewed officials from SBA and the 11 SBIR participating agencies about SBIR, including SBA's rule implementing the NDAA's amendments to

SBIR's eligibility criteria and their decisions on whether to submit a written determination. We also interviewed officials from three industry associations representing small businesses or venture capital firms, a researcher from the National Academy of Sciences, a venture capital firm, and a consultant about SBIR, including the potential impact of expanding SBIR to include majority-owned portfolio companies and the potential interest of such companies in SBIR. We selected these organizations and individuals to interview based on our review of the congressional record and SBIR hearings and our discussions with SBA and participating agencies. In addition, we interviewed all four majority-owned portfolio companies that received SBIR awards from one of the two participating agencies that submitted a written determination, as of May 2014, when we made our initial data request. Finally, we interviewed four majority-owned portfolio companies that registered with SBA but had not yet applied for an SBIR award to obtain information on their experiences with SBIR. We selected the four companies based on their percent ownership by venture capital and area of primary research.

We conducted this performance audit from February 2014 to November 2014 in accordance with generally accepted government auditing standards. Those standards require that we plan and perform the audit to obtain sufficient, appropriate evidence to provide a reasonable basis for our findings and conclusions based on our audit objectives. We believe that the evidence obtained provides a reasonable basis for our findings and conclusions based on our audit objectives.

BACKGROUND

Federal agencies with a budget in excess of $100 million for extramural R&D are required to establish and operate an SBIR program.[8] In fiscal year 2013, agencies participating in the SBIR program were required to spend at least 2.7 percent of their extramural R&D budgets on SBIR awards. Currently, 11 agencies participate in the SBIR program: the Departments of Agriculture, Commerce, Defense, Education, Energy (DOE), Health and Human Services (HHS), Homeland Security, and Transportation, and the Environmental Protection Agency, National Aeronautics and Space Administration, and National Science Foundation. Although each agency manages its own program, SBA plays a central administrative and oversight role.

The SBIR program includes the following three phases:

- In phase I, agencies make awards to small businesses to determine the scientific and technical merit and feasibility of ideas that appear to have commercial potential. Phase I awards normally do not exceed $150,000. For SBIR, work in phase I generally lasts 6 to 9 months.
- In phase II, small businesses with phase I projects that demonstrate scientific and technical merit and feasibility, in addition to commercial potential, may compete for awards of up to $1 million to continue the R&D for an additional period, normally not to exceed 2 years.[9]
- Phase III is for small businesses to pursue commercialization of technology developed in prior phases. Phase III work derives from, extends, or completes an effort made under prior phases, but it is funded by sources other than the SBIR program. In this phase, small businesses are expected to raise additional funds from private investors, the capital markets, or from non-SBIR funding sources within the government. While SBIR funding cannot be used for phase III, agencies can participate in phase III by, for example, purchasing the technology developed in prior phases.[10]

SBA's Office of Investment and Innovation is responsible for overseeing and coordinating the participating agencies' efforts for the SBIR program by setting overarching policy and issuing policy directives, collecting program data, reviewing agency progress, and reporting annually to Congress, among other responsibilities. As part of its oversight and coordination role, SBA issued an updated SBIR Policy Directive in January 2014.[11] The directive explains and outlines requirements for agencies' implementation of the SBIR program. The policy directive includes information on program eligibility, proposal requirements, terms of agreement for SBIR awards, and responsibilities of SBA and participating agencies for the program.

Each participating agency must manage its SBIR program in accordance with program laws, regulations, and the policy directive. Each participating agency has considerable flexibility to design and manage the specifics of the program, such as determining research topics, selecting award recipients, and administering funding agreements. All of the agencies follow the same general process to obtain proposals from and make awards to small businesses for the SBIR program. At least annually, each participating agency issues a solicitation requesting proposals for projects in topic areas determined by the agency. Each agency uses its own process to review proposals and determine which proposals should receive awards. Also, each agency determines whether the funding for awards will be provided as grants or contracts.

The NDAA amended the SBIR program to permit agencies to allow participation by majority-owned portfolio companies, and the act imposes certain requirements on such companies and participating agencies. As discussed, before awarding SBIR funds to majority-owned portfolio companies, agencies must submit to SBA and Congress a written determination. The NDAA also imposes statutory caps on the percentage of participating agencies' SBIR funds that may be awarded to majority-owned portfolio companies—25 percent for NIH, DOE, and the National Science Foundation, and 15 percent for the other participating agencies. If an agency awards more than the percentage of the funds set by the statutory cap, the agency must transfer any amount in excess of the cap from its non-SBIR R&D funds to the agency's SBIR funds. Additionally, agencies may not use investment of venture capital or investment from hedge funds or private equity firms as a criterion for the award of contracts under the SBIR program. The NDAA also requires all majority-owned portfolio companies to register as such with SBA and indicate their majority-owned portfolio company status in any SBIR application. Finally, the NDAA required SBA to update the SBIR Policy Directive to conform to NDAA amendments, such as majority-owned portfolio company participation in the program.

CHANGE IN SBIR PROGRAM ELIGIBILITY HAS HAD LITTLE IMPACT

In 2013, HHS and DOE each provided a written determination to SBA and Congress prior to making SBIR awards to majority-owned portfolio companies.[12] More specifically, HHS's written determination covered one of its four SBIR subunits, NIH, which according to agency officials accounted for about 98 percent of the HHS SBIR funds in fiscal year 2013.[13] DOE's written determination covered one of its two subunits that participate in SBIR, the Advanced Research Projects Agency-Energy (ARPA-E), which according to agency officials accounted for about 4 percent of the DOE SBIR awards in fiscal year 2013.[14] NIH and ARPA-E officials told us that by opening their SBIR programs to majority-owned portfolio companies, they would help ensure that they received the highest quality applications with the best scientific research, regardless of whether a small business has venture capital support. In addition, NIH officials said that allowing majority-owned portfolio companies to participate in SBIR would increase the flexibility for SBIR companies to seek additional investment sources that would help support SBIR's goal of commercialization. As part of its written determination, NIH cited a 2009 National Academy of Sciences study, whose findings suggested that the most commercially promising companies were those that were repeatedly selected by both NIH for their promising technologies and by venture capital investors for their commercial potential.[15]

A few majority-owned portfolio companies have participated in SBIR since SBA's final rule implementing the NDAA's changes to the SBIR eligibility requirements took effect in January 2013. NIH and ARPA-E opened their solicitations to majority-owned portfolio companies in April 2013 and June 2013, respectively.[16] From October 2012 (start of fiscal year 2013) through September 2014 (end of fiscal year 2014), these subunits issued 68 SBIR solicitations, of which 56 were open to majority-owned portfolio companies.[17] More specifically, ARPA-E opened one solicitation to majority-owned portfolio companies that covered both fiscal years 2013 and 2014, and NIH opened 55 solicitations to majority-owned portfolio companies during fiscal years 2013 and 2014.

In response to their SBIR solicitations open to majority-owned portfolio companies, ARPA-E and NIH received a small number of applications from such companies and made several awards to them. As shown in figure 1, majority-owned portfolio company applications received and awards made comprise less than 1 percent of NIH's and ARPA-E's SBIR applications and awards. ARPA-E and NIH collectively received a total of 20 applications from majority-owned portfolio companies in fiscal years 2013 and 2014, compared to 11,906 applications from applicants that were not majority-owned portfolio companies. In addition, these two subunits made 12 SBIR awards to majority-owned portfolio companies. Specifically, ARPA-E made 2 SBIR awards to two majority-owned portfolio companies, which included one phase I award to one company and another award that included multiple phases to another company.[18] In addition, NIH made awards to 10 majority-owned portfolio companies that consisted of 7 phase I awards, and 3 phase II awards.

ARPA-E and NIH officials told us that it was too early for them to evaluate the impact of including majority-owned portfolio companies in their SBIR programs—such as how the composition of applicants and awardees might change—but said that this change has not created any administrative problems or burdens. The officials from both subunits told us that they did not encounter any challenges completing their written determination, updating their data systems, and monitoring their compliance with the statutory cap.

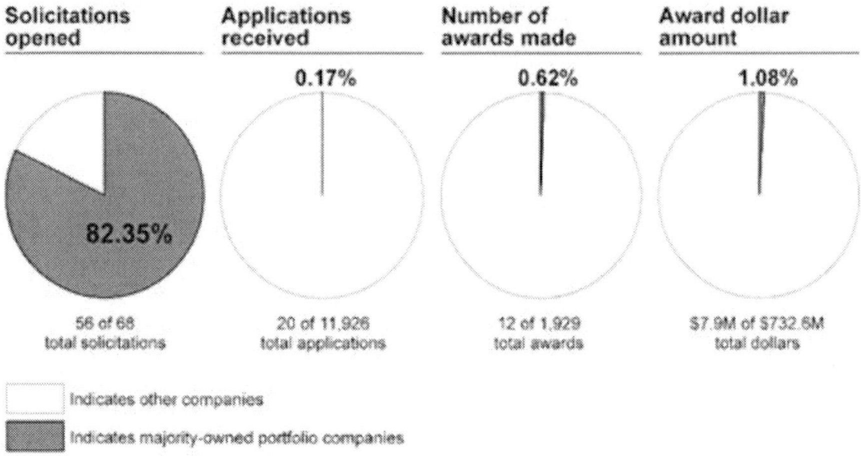

Source: GAO analysis of National institute of Health and Advanced Research Projects Agency-Energy data. |GAO-15-68.

Note: Solicitations opened to majority-owned portfolio companies also were open to other eligible companies, while some solicitations opened to other eligible companies were not open to majority-owned portfolio companies.

Figure 1. Summary of Majority-Owned Portfolio Company and Other Company SBIR Participation in ARPA-E and NIH in Fiscal Years 2013 and 2014.

ARPA-E is the only DOE subunit using the majority-owned portfolio company funding option, and so under the statute, it could potentially award up to 25 percent of DOE's total SBIR funds to such companies. DOE and ARPA-E officials told us they administer the cap at the subunit level—meaning that ARPA-E may award up to 25 percent of its SBIR funds to majority-owned portfolio companies. According to ARPA-E officials, for fiscal year 2013, ARPA-E made approximately $7 million in SBIR awards, of which about $1.7 million (or nearly 25 percent of its SBIR awards) was awarded to a majority-owned portfolio company. For fiscal year 2014, ARPA-E officials told us that the subunit was just under its fiscal year 2014 internal cap of 25 percent of its SBIR funds. NIH consists of 27 institutes and centers, and 24 institutes and centers participate in the SBIR program. NIH officials told us that they are applying the 25 percent statutory cap to each institute and center. Officials told us that NIH as a whole was not close to its 25 percent statutory cap for fiscal year 2014, having awarded about $4.5 million to majority-owned portfolio companies, which was about 0.68 percent of the agency's $663 million SBIR award obligations for fiscal year 2014.

Representatives from four majority-owned portfolio companies that received NIH or ARPA-E SBIR awards told us that their SBIR awards have allowed them to conduct new research that would otherwise not have been undertaken. For example, one representative said that venture capital firms often view R&D for new projects that they have not funded as a distraction from a firm's focus on commercializing a product and his company would have faced difficulties conducting its research without SBIR funding. Additionally, representatives from these four companies we interviewed told us that receiving an SBIR award can be viewed as validating the merit of their research, which can attract additional venture capital funds and improve their ability to commercialize their SBIR research. The representatives said they did not encounter any significant problems applying for SBIR awards or receiving SBIR funds.[19]

Agencies Cite Various Reasons for Electing Not to Participate Most Participating Agencies Have Not Exercised Their Authority to Open Their SBIR Programs to Majority-Owned Portfolio Companies

Nine agencies have chosen not to open their SBIR programs to majority-owned portfolio companies, and therefore have not submitted a written determination to do so. Specifically, as of September 2014, the Departments of Agriculture, Commerce, Defense, Education, Homeland Security, and Transportation; Environmental Protection Agency; National Aeronautics and Space Administration; and National Science Foundation had not taken such action. The awards from these nine agencies, along with those from the programs within HHS and DOE that chose not to allow portfolio company participation, accounted for about 71 percent of the nearly $2.1 billion SBIR awards in fiscal year 2013.

Officials from the nine agencies generally told us that the decision not to open their programs to portfolio companies was made at the program level for each agency. They said the decision was based largely on internal discussions among SBIR staff and, in some cases, staff from other departments, and did not involve any formal analyses. The explanations that agency officials told us during our interviews for not submitting a written determination were as follows.

- *Department of Agriculture:* An official said that the Department viewed other administrative changes mandated under the NDAA as a higher priority, given the low level of interest from majority-owned portfolio companies in their SBIR program.[20] They also said that the agency's SBIR awards are smaller than other agencies' awards, such as the Department of Defense and NIH, and majority-owned portfolio companies are most interested in the agencies offering the largest awards.
- *Department of Commerce:* Officials said that they viewed other administrative changes—such as changes to its proposal evaluation process—as a higher priority than the majority-owned portfolio company funding option. Also, the officials stated that they were concerned that adopting the funding option could lead to a significant increase in the number of proposals, and if that happened, there would be a corresponding decrease in the percentage of proposals receiving SBIR awards, given the small size of the agency's SBIR program compared to other federal agencies' programs. Officials said the agency will be in a better position to understand the possible results of implementing the funding option and can better prepare the agency's SBIR program once they see how the change impacts other agencies.
- *Department of Defense:* Officials said that the agency does not have the evidence to support that allowing majority-owned portfolio companies in SBIR would achieve the outcomes required to be specified in the written determination, such as substantially contributing to the agency's mission. Officials also said they do not believe that there is much interest from venture capitalists in investing in firms that would be eligible to receive SBIR phase I and phase II awards.
- *Department of Education:* Officials said that the agency does not have an official reason why it did not elect to exercise the majority-owned portfolio company funding option. They told us that majority-owned portfolio companies have not approached the agency about its SBIR program for many years; thus, they did not see a reason to use the funding option.
- *Department of Homeland Security:* Officials said that they viewed other administrative changes mandated under the NDAA as a higher priority and that they had not yet conducted the research to provide

evidence that majority-owned portfolio companies would achieve the outcomes required to be outlined in the written determination requirement. The officials also said that the agency's SBIR topics are narrowly defined to respond to the technology gaps in the agency's operating components. As a result, they said there is a lack of scalability for the general market; therefore, their SBIR program does not generally draw attention and interest from the venture capital community. The officials said, however, that they are working to gain greater partnership interest with industry to make these technologies more desirable in commercial markets, which may gain interest from the venture capital community.

- *Department of Transportation:* Officials said that they were concerned that opting in could increase the number of SBIR applications, which would impact the agency's resources and ability to meet the SBIR award timelines. Officials need additional time to consider the potential impact of opening its SBIR program up to majority-owned portfolio companies and how the agency can best manage that impact. The officials also told us that the agency has not had any majority-owned portfolio companies express interest in its SBIR program, as of March 2014 when we held our interview, so officials did not see a pressing reason to immediately opt-in.
- *Environmental Protection Agency:* Officials said that the agency currently receives many SBIR proposals and can only fund a small number of them; thus, they did not need a new pool of applicants for the program, especially considering that majority-owned portfolio company applicants could be small businesses that have better funding since they are more than 50 percent owned by venture capital operating companies, hedge funds, or private equity firms. They told us that the agency would have to do a rigorous and detailed analysis to complete the written determination and do not know what factors they would consider to make the decision.
- National Aeronautics and Space Administration: Officials said that the agency does not have statistically verifiable evidence to support that allowing majority-owned portfolio companies into its SBIR program would achieve the outcomes required to be enumerated in the written determination. The officials told us that firms with majority ownership by venture capital may have a low level of interest in their SBIR program because of the long length of the research projects in this field.

- National Science Foundation: Officials told us that they generally did not believe they have the evidence to support that allowing majority-owned portfolio companies into its SBIR program would achieve the outcomes required to be enumerated in the written determination. For example, they told us that the agency already has an extensive relationship with the venture capital and the angel investor communities through SBIR, so they did not believe opting in would address the need for publicly funded research. Officials also said that the National Science Foundation is not an acquisition agency, so such action would not substantially contribute to the agency's mission. Finally, the officials told us that majority-owned portfolio companies have not expressed interest in the agency's SBIR program.
- All but one of the agencies told us they may reevaluate their decisions in the future, but generally did not have any specific plans for doing so. For example, officials at the Department of Transportation said they could not determine the specific factors they would review when they reconsider their decision, because it is too early in the implementation of the provision to see what factors would affect their decision. Officials at the Departments of Commerce and Transportation and the Environmental Protection Agency said they are waiting to see how the change in eligibility affected ARPA-E and NIH before they reconsider implementing the change at their own agency. SBIR program managers at the Department of Defense told us that they do not plan to reevaluate their decision as they do not see a benefit in allowing such companies to participate in SBIR.

Although some agencies considered the potential interest of majority-owned portfolio companies in their SBIR programs in making their decision on whether to submit a written determination, gauging such interest can be difficult. Information on the degree of venture capital ownership in a company is confidential and proprietary and not reported to any public databases. Most participating agencies historically have not tracked whether applicants or awardees have venture capital funding. As specified in the Policy Directive, agencies must require awardees to certify their eligibility at the time of the award, including certifying whether they are majority-owned portfolio companies, but the agencies generally do not collect any detailed information on an applicant's or awardee's ownership.[21] In 2009, the National Academy of Sciences reported that between 4 percent and 12 percent of the small businesses that won SBIR phase II awards from NIH between 1992 and 2002 were majority-owned portfolio companies; however, the study did not provide any information on majority-owned portfolio company participation at other agencies.[22] Additionally, in 2006, we found that approximately 17 percent of NIH awards and 7 percent of the Department of Defense awards went to small businesses that had venture capital investment, but this study did not differentiate between majority- and minority-owned portfolio companies.[23] SBA requires companies that apply for SBIR awards to register on its Company Registry and indicate whether they are majority-owned portfolio companies. As of September 2014, 153 majority-owned portfolio companies had registered with SBA and they accounted for less than 2 percent of the total number of small businesses registered with SBA.

Data are not publicly available to gauge the interest of majority-owned portfolio companies in the SBIR program, but industry representatives indicated that such companies may not be aware of the program change and could be interested in the program. A representative from an industry association representing venture capitalists told us that because majority-owned portfolio companies have been unable to participate in SBIR previously, many of them may not be aware of the change in program eligibility at some agencies. Representatives from three industry associations told us that venture capital investment in R&D has diminished over the past decade, and representatives from two of these associations said the SBIR program is very attractive to majority-owned portfolio companies because it allows them to apply for SBIR awards to fill the gap in R&D funding. We selected a nonprobability sample and interviewed four of the majority-owned portfolio companies that registered with SBA but had not applied for an SBIR award.[24] Officials from three of the firms said that they had not seen an SBIR solicitation that fit their type of research, while officials from one firm said they still were working to determine their eligibility for the program.

Some Agencies Are Interpreting the Written Determination as a Potentially Stringent Requirement

In our discussions with the nine participating agencies that did not open their SBIR programs to majority-owned portfolio companies, we found that six agencies viewed the written determination as a potentially stringent requirement. The NDAA requires agencies to submit a written determination to SBA and Congress at least 30 days before making SBIR awards to majority-owned portfolio companies. Pursuant to the NDAA, the written determination must explain how the use of the authority to allow such companies to participate in the SBIR program will (1) induce additional venture capital, hedge fund, or private equity firm funding of small business innovations; (2) substantially contribute to the agency's mission; (3) demonstrate a need for public research; and (4) otherwise fulfill the capital needs of small business concerns for additional financing for SBIR projects.[25] According to SBA officials, the written determination is a notification letter, serving to inform SBA and Congress of the agency's plans. SBA reviews these determinations, but it does not approve or deny them.

In their written determinations, NIH and DOE supported their explanations largely based on their experience with SBIR and general knowledge about majority-owned portfolio companies. They did not conduct any independent research on majority-owned portfolio companies, which is not specifically required by NDAA, but NIH cited a 2009 study conducted by the National Academy of Sciences on its SBIR program. 26 Officials from NIH and DOE told us that they did not have any issues preparing their written determination letter. In contrast, several other agencies viewed the written determination as a potentially stringent requirement that necessitated independent analyses. For example, officials at the National Aeronautics and Space Administration said that they did not have statistically verifiable evidence that majority-owned portfolio company participation would produce the written determination outcomes, and officials at the Department of Homeland Security told us they did not have enough data to show that such companies would be interested in its SBIR program. Likewise, officials at the Department of Defense told us that they generally did not have the evidence to support that allowing majority-owned portfolio companies would achieve the written determination outcomes, and officials at the Environmental Protection Agency said they would need to conduct a rigorous analysis to make the determination. Officials at the National Science Foundation told us that they did not have the evidence to support the written determination outcomes, while a program manager at the Department of Agriculture said that the requirement is an administrative burden, and if the barrier was lower, the agency might allow majority-owned portfolio companies to participate in its SBIR program. The other three participating agencies did not discuss any issues with the written determination requirement.

As required under the NDAA, SBA updated its SBIR Policy Directive in response to the NDAA's reauthorization of, and amendments to, the SBIR program. Specifically, SBA revised the policy directive to include, among other things, the written determination requirement. The policy directive essentially uses the same language as the NDAA and does not provide any specific guidance on what evidence participating agencies may need to consider to comply with the written determination requirement. SBA officials told us that the agency's role with the participating agencies is to provide a forum for all of the program managers to discuss SBIR, including the new majority-owned portfolio company funding option. The officials said that SBA has not tried to encourage or discourage agencies to adopt the new funding option. According to the SBA officials, SBA meets routinely with SBIR program managers, and no agency has raised concerns about the written determination requirement. Similarly, most SBIR program managers told us that they have not had any discussions with SBA about whether they should allow majority-owned portfolio companies to participate in their programs, and that they meet monthly with SBA and periodically communicate with SBA officials through telephone calls and e-mail.

In its proposed and final rules implementing the NDAA provisions, SBA stated one potential benefit of the rule is to provide more businesses with access to the SBIR program, which would increase competition and the quality of proposals and spur innovation.[27] Some agencies may be viewing the evidence required for the written determination differently, and may view the written determination as a potentially stringent requirement. SBA officials told us that they are not statutorily required to advise participating agencies on what evidence they should consider in their written determination. Nonetheless, by providing additional guidance on the requirement, SBA could better inform the agencies about the evidence that they may consider in the written determination to explain how making SBIR awards to majority-owned portfolio companies will, among other things, induce additional venture capital or similar funding of small business innovation and substantially contribute to the agency's mission. As a result, agencies would be better positioned if they choose to reevaluate their decision on whether to use the majority-owned portfolio company funding option, which most agencies plan to do in the future.

CONCLUSION

The effect of the NDAA provisions allowing SBIR participating agencies to permit majority-owned portfolio companies to participate in their SBIR programs has been limited, because the majority of participating agencies have opted not to open their SBIR programs to such companies. As the administrator of the SBIR program, SBA updated its SBIR Policy Directive to conform to the amendments made by NDAA but did not provide specific guidance on the evidence participating agencies may consider to comply with the written determination requirement. At the same time, NIH and DOE were able to complete their written determinations, which SBA reviewed. Although the other participating agencies have not asked SBA for additional guidance on the written determination requirement, SBA also has not discussed the issue with them, such as at their monthly meetings or as part of other outreach efforts. As a result, agencies may be uncertain about the evidence necessary to support the written determination and may view the determination as a potentially stringent requirement. SBA is not responsible for encouraging or discouraging agencies to use the new authority and expand eligibility to include majority-owned portfolio companies, but SBA, as the program administrator, could be missing an opportunity to help agencies better understand the evidence required for the written determination, which could inform agencies' decisions about whether to expand their program.

Recommendations for Executive Action

To help ensure that participating agencies understand the requirements of the NDAA provisions applicable to allowing majority-owned portfolio companies to apply for SBIR awards, we recommend that the Administrator of the Small Business Administration discuss the evidence required for the written determination with the participating agencies, such as at their monthly meeting or as part of another outreach effort, and, if needed, and in consultation with the participating agencies, amend its SBIR Policy Directive to provide additional guidance.

Agency Comments and Our Evaluation

We provided a draft of this report to the Secretaries of Health and Human Services, Agriculture, Commerce, Defense, Education, Energy, Homeland Security, and Transportation; the Administrators of the Small Business Administration, the Environmental Protection Agency, and the National

Aeronautics and Space Administration; and the Director of the National Science Foundation for review and comment. SBA, the Departments of Health and Human Services, Homeland Security, and Transportation, and the National Science Foundation provided technical comments, which we incorporated, as appropriate.

SBA SBIR officials provided comments on a draft of our report through an e-mail from their GAO liaison on October 29, 2014. In these comments, SBA stated that it intends to discuss the written determination with SBIR program managers at a future program managers meeting. However, SBA stated it is not required by statute to advise participating agencies on what evidence they should consider in their written determination. We clarified this in the final report. Additionally, SBA said the SBIR Policy Directive contains guidance on what the written determination must include, and participating agencies have not requested more detailed guidance on the written determination.

We maintain that our findings and recommendation on providing guidance on the written determination are appropriate. As we noted above, SBA may not be required by law to advise agencies on the evidence they may use to support the written determination, but SBA could be missing an opportunity as program administrator to help agencies make a more informed decision about whether to expand their SBIR program. SBA notes that the SBIR Policy Directive contains guidance on the written determination and no agency has requested more detailed guidance.

However, we found that participating agencies were viewing the evidence required for the written determination differently. Specifically, six of the nine agencies that have not opted to expand their program viewed the written determination as potentially stringent, possibly requiring independent analysis, while the two agencies that opted to expand their program viewed the written determination as a less stringent requirement. Our recommendation addresses this issue, and SBA's plan to discuss the written determination with SBIR program managers at a future meeting is consistent with our recommendation.

Cindy Brown Barnes,
Director,
Financial Markets
and Community Investment

End Notes

[1] Small Business Innovation Development Act of 1982, Pub. L. No. 97-219, 96 Stat. 217 (1982).

[2] See Committee for Capitalizing on Science, Technology, and Innovation: An Assessment of the Small Business Innovation Research Program, National Academy of Sciences, Venture Funding and the NIH SBIR Program, Charles W. Wessner, ed. (The National Academies Press, Washington, D.C.: 2009).

[3] To qualify for SBIR awards, small businesses must meet certain SBIR ownership and other eligibility criteria, such as being more than 50 percent owned by individuals who are U.S. citizens or permanent resident aliens. 13 C.F.R. § 121.702(a). In 2003, an SBA administrative law judge issued a decision stating that venture capital firms could not be considered "individuals" for the purpose of satisfying the ownership criteria for the program. Size Appeal of Cognetix, Inc., SBA No. SIZ-4560 (2003).

[4] Pub. L. No. 112-81, Tit. LI, 125 Stat. 1298, 1823 (2011). The NDAA also reauthorized the Small Business Technology Transfer program, which was initiated in 1992 for the purpose of stimulating a partnership of ideas and technologies between innovative small businesses and research institutions through federally funded R&D. Research institutions include nonprofit colleges or universities and federally funded research and development centers. The Small Business Technology Transfer program is outside the scope of our review because SBA's final rule does not permit majority-owned portfolio companies to participate in that program.

[5] 77 Fed. Reg. 76,215 (Dec. 27, 2012).

[6] Specifically, an agency must submit a written determination to the Senate Committee on Small Business and Entrepreneurship, the House Committee on Small Business, and the House Committee on Science, Space, and Technology at least 30 calendar days before it begins making awards to majority-owned portfolio companies. The written determination must include four elements that explain how allowing majority-owned portfolio companies will: (1) induce additional venture capital, hedge fund, or private equity firm funding of small business innovations; (2) substantially contribute to the mission of the federal agency; (3) demonstrate a need for public research; and (4) otherwise fulfill the capital needs of small business concerns for additional financing for SBIR projects. 15 U.S.C. § 638(dd)(1)-(2).

[7] Pub. L. No. 112-81, § 5142 (codified at 15 U.S.C. § 638a). The NDAA also includes provisions that require the participating agencies, in consultation with SBA, to commission a study by the National Academy of Sciences to review the SBIR program, including the value and quality of the research, economic and non-economic benefits, and jobs created. Additionally, the National Research Council has completed studies on the SBIR program under previous mandates, including a study on the impact of venture funding at the National Institutes of Health.

[8] Agencies' R&D programs generally include funding for two types of R&D: intramural and extramural. Intramural R&D is conducted by employees of a federal agency in or through government-owned, government-operated facilities. Extramural R&D generally is conducted by nonfederal employees outside of federal facilities. Agencies are required to calculate their extramural R&D budget by subtracting amounts obligated for intramural R&D from total obligations for R&D.

[9] Agencies may award a second, sequential, phase II award to continue a phase II project.

[10] For examples of how one agency has used phase III awards, see GAO, Small Business Innovation Research: DOD's Program Supports Weapon Systems, but Lacks

Comprehensive Data on Technology Transition Outcomes, GAO-14-96 (Washington, D.C.: Dec. 20, 2013).

[11] SBA, Office of Investment and Innovation, Small Business Innovation Research (SBIR) Program Policy Directive, www.sbir.gov (Washington, D.C.: Feb. 24, 2014).

[12] HHS submitted its determination letter on March 7, 2013, and DOE submitted its letter on August 29, 2013.

[13] HHS has four subunits that participate in SBIR: (1) the Administration for Children and Families; (2) Centers for Disease Control and Prevention (CDC); (3) Food and Drug Administration; and (4) NIH. NIH's mission is to improve the health of humans. NIH is the largest source of funding for medical research in the world, and is made up of 27 institutes and centers that each has a specific research agenda. More than 80 percent of NIH's budget goes to more than 300,000 research personnel at over 2,500 universities and research institutions. CDC submitted its written determination letter on July 30, 2014, and expects to begin making awards during fiscal year 2015. Thus, CDC is not part of this review.

[14] ARPA-E's mission is to reduce U.S. dependence on energy imports, reduce U.S. energy-related emissions, improve energy efficiency across all sectors of the U.S. economy, and ensure that the United States maintains a technological lead in the development and deployment of advanced energy technologies.

[15] Committee for Capitalizing on Science, Technology, and Innovation: Venture Funding and the NIH SBIR Program, 2009.

[16] NIH and ARPA-E, along with SBA, have used various means to inform majority-owned portfolio companies about the eligibility changes in their SBIR programs. On its SBIR website, SBA identifies the agencies that have opted to allow majority-owned portfolio companies to participate in their SBIR programs. According to NIH and ARPA-E officials, their solicitations now specifically state that majority-owned portfolio companies may participate. Further, NIH officials told us that they e-mailed approximately 15,000 subscribers, announcing the change in their agency's eligibility requirements.

[17] NIH and ARPA-E provided us with data for fiscal year 2014 before subjecting the data to their review process in preparation of their annual reports, which are typically due to SBA by March 15, 2015. As a result, our data may differ slightly from the data in their annual reports.

[18] ARPA-E can make two types of combined awards: (1) combined phase I/phase II awards may be funded up to $1,725,000 with a period of performance up to 36 months and (2) combined phase I/phase II/phase II sequential awards may be funded up to $3,225,000 and may have a period of performance up to 48 months.

[19] These comments were consistent with findings from our prior work in which we found that venture capital firms are not typically interested in funding research covered by the SBIR program because such investments are high risk and typically have slow returns. We also reported that receiving an SBIR award served as a "stamp of approval" to venture capital firms, which means that venture capital firms may be more willing to fund projects. Our 2012 report summarized three reasons that explain why venture capital is not interested in this type of research: (1) Private investors consider these projects too high risk for investment, even for concepts with promising technological potential; (2) venture capital firms seek more rapid returns on investment and closely analyze a project's potential return on investment over time, a factor that influences their decisions to invest in projects that are in later stages of development and closer to commercialization; (3) venture capital firms may not be comfortable investing in new energy technologies, noting the historical lack of successful venture capital investments in these types of projects. See GAO, Department of

Energy: Advanced Research Projects Agency-Energy Could Benefit from Information on Applicants' Prior Funding, GAO-12-112 (Washington, D.C.: Jan.13, 2012).

[20] In addition to changing SBIR's eligibility requirements, the NDAA made several key changes to the SBIR award process, SBIR program administration, and fraud, waste, and abuse prevention procedures. For the award process, the act requires agencies to make SBIR award decisions within a certain amount of time after the close of the solicitation. For program administration, the act's new requirements include creating and setting forth the policies for the new pilot program that permits agencies to use SBIR money for administration of the SBIR program and setting forth the new reporting and data collection requirements. Finally, the act included amendments relating to fraud, waste, and abuse, such as requiring small businesses to certify they are meeting the program's requirements during the life cycle of the funding agreement.

[21] ARPA-E reported that applicants are required to report venture capital funding on the "ARPA-E Other Sources of Funding" form. In addition, awardees report post-award venture capital funding in their quarterly reports submitted to ARPA-E.

[22] Committee for Capitalizing on Science, Technology, and Innovation: Venture Funding and the NIH SBIR Program, 2009.

[23] GAO, Small Business Innovation Research: Information on Awards Made by NIH and DOD in Fiscal Years 2001 through 2004, GAO-06-565 (Washington, D.C.: Apr. 14, 2006).

[24] We used a multi-step procedure to select companies that ranged in percentage of ownership by venture capital, and were not in the biotechnology or electricity industries, as these areas were covered by NIH and DOE solicitations.

[25] 15 U.S.C. § 638(dd)(2).

[26] Committee for Capitalizing on Science, Technology, and Innovation: Venture Funding and the NIH SBIR Program, 2009.

[27] 77 Fed. Reg. 75,215, 76,223 (Dec. 27, 2012) (final rule); 77 Fed. Reg. 28,520, 28,526 (May 15, 2012) (proposed rule).

In: Small Business Innovation Research ... ISBN: 978-1-63482-532-0
Editor: Thomas H. Chavez

Chapter 4

SBIR/STTR OUTREACH*

SBIR/STTR Interagency Policy Committee

1. EXECUTIVE SUMMARY

Purpose. This report responds to the statutory requirement in the Small Business Innovation Research (SBIR)/Small Business Technology Transfer (STTR) Reauthorization Act of 2011 (Reauthorization Act), which was included in the 2012 Defense Authorization Act (P.L.112-81)[1]. The Reauthorization Act requires the Interagency Policy Committee (IPC) to review the issues regarding increased outreach about the SBIR/STTR programs to socially disadvantaged small businesses and small business that are located in underserved communities.

Key Issues. The key issues that need to be addressed include the following: (1) determine cost effective and innovativeways to conduct outreach and education to socially disadvantaged small businesses and small businesses that are located in underserved communities: (2) ensure accurate data reporting on awards to such small businesses; and (3) determine the effectiveness of the outreach efforts by continuing to review program participation bythese *Key Recommendations.*

---

* This is an edited, reformatted and augmented version of a report to Congress issued by the Small Business Innovation Research (SBIR) and Small Business Technology Transfer (STTR) Program Interagency Policy Committee, September 15, 2014.

[1] http://www.gpo.gov/fdsys/pkg/PLAW-112publ81/pdf/PLAW-112publ81.pdf.

- Unify the outreach effort across the SBIR/STTR Programs for cost effectiveness.
- Implement new methods of communication and outreach tools using state, regional and local agencies and trade groups.
- Develop a new series of narratives around the programs highlighting its impact to these communities and its correlation around economic development and prosperity.
- Monitor the outreach efforts and compare to program participation by socially disadvantaged small businesses and small businesses that are located in underserved communities.

2. INTRODUCTION

The Reauthorization Actcreated the SBIR/STTR Interagency Policy Committee (IPC)to review certain issues and make policy recommendations to Congress on ways to improve program effectiveness and efficiency. The IPC is co-chaired by the White House Office of Science and Technology Policy (OSTP) and the Small Business Administration (SBA) and includes representatives from Federal agencies that participate in the SBIR or STTR programs to review certain issues and make policy recommendations to Congress on ways to improve program effectiveness and efficiency.

a. Background

Congress established the SBIR program in 1982 and the STTR program ten years later to assist small business concerns (SBCs) in obtaining Federal research and development (R&D) funds to build a strong economy and support technological innovation as discussed below:

- *SBIR*: Requires federal agencies with extramural R&D budgets over $100 million to set aside a percentage of their annual extramural R&D budget for small businesses. This percentage was 2.5% prior to FY 2012, and increased to 2.6% in FY 2012 and will continue to increase by 0.1% each year until it reaches a base requirement of 3.2% in FY 2017.

- *STTR*: Modeled after the SBIR program, STTR requires federal agencies with extramural budgets exceeding $1 billion to set aside a percentage of their annual extramural R&D budget for SBCs that work in cooperation with universities, federally funded research and development centers, and other non-profit scientific and educational institutions. This percentage was 0.3% in FYs 2004-2011, and legislation increased this minimum to 0.35% for FYs 2012 and 2013, with continued increases through 2016. The goal is to facilitate the transfer of technology and research from these institutions to commercial use and encourage innovation.

b. Program Objectives

The Small Business Act indicates that "assistance" provided by the SBIR and STTR programs "be given to small-business concerns to enable them to undertake and obtain the benefits of research and development in order *to maintain and strengthen the competitive free enterprise system and the national economy.*"[2]

The SBIR Policy Directive identifies the following primary objectives:

- Stimulate technological innovation;
- Meet Federal R&D needs while strengthening the role of small businesses;
- Foster and encourage participation in innovation and entrepreneurship by socially and economically disadvantaged persons and women-owned small businesses; and
- Increase private-sector commercialization of innovations derived from Federal research and development funding.

In addition, the Reauthorization Act added several new initiatives including:

- Permitting agencies to direct some SBIR funds to firms that are owned by multiple venture capital operating companies (VCOCs), hedge funds, or private equity firms;
- Allowing agencies to provide one additional Phase II award to small businesses to extend a study;

[2] 15 U.S.C. § 638(a).

- Providing for a Commercialization Readiness Program at DOD and pilot programs at civilian agencies;
- Reducing processing times;
- Allowing agencies to use 3% of program budgets for administration and oversight;
- Reducing vulnerability of Fraud Waste and Abuse (FWA);
- Improving program evaluation; and
- Increasing commercialization and outreach.

In general, the SBIR/STTR Reauthorization Act underscored the need for improved commercialization, outreach, and program evaluation. In December 2013 the lead SBIR/STTR Program Managers in conjunction with White House Office of Science & Technology Policy (OSTP) facilitated the creation of five sub-working groups that fall under the "Fueling Small Business Innovation" component of the President's Lab to Market Commercialization Agenda.

These five groups are:

1. Outreach & Communications
2. Commercialization Pathways
3. Awards Efficiency & Efficacy
4. Databases & Interagency Exchange of Information
5. Asset Mapping

Going forward these five groups are tasked with various short-term and long-term projects that fall within scope of various facets covered in the President's Lab to Market Commercialization Agenda. They will seek to identify issues, challenges, and provide further recommendations for consideration amongst the various SBIR/STTR program managers as well for SBA and OSTP consideration as it relates to the SBIR/STTR program.

c. SBIR/STTR Process

To bring technology from ideas to commercialization, both programs utilize a three phase approach:

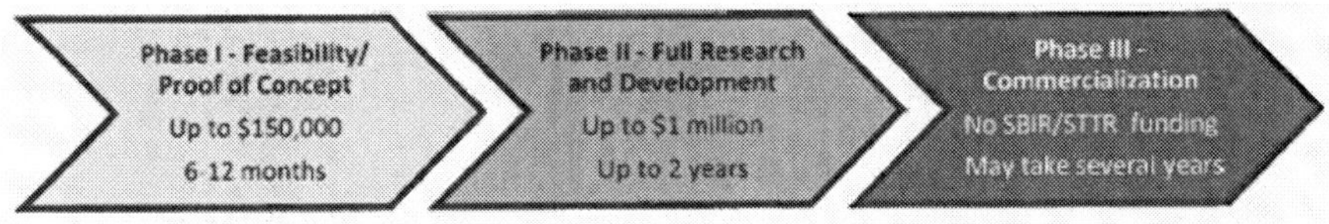

- *Phase I – Feasibility/Proof of Concept.* Using a competitive process, federal agencies award up to $150,000 to a small business to perform research/research and development (R/R&D) for up to 6-12 months on a specific topic in order to establish its technical merit, feasibility, and commercial potential. During this phase, federal agencies assess both the performance of the small business and the potential of the technology prior to providing further Federal support in Phase II.
- *Phase II – Full Research and Development.* Based on the results achieved in Phase I, federal agencies will decide whether to continue R/R&D efforts into Phase II based on the scientific, technical, and commercial merit and feasibility of the idea. If the federal agency decides to continue into Phase II, they will award up to $1 million to the small business to continue R/R&D efforts for up to 2 years.
- *Phase III - Commercialization.* No specific SBIR funding is associated with Phase III, however, some agencies may include follow-on non-SBIR funded R&D or production contracts for products or services intended for use by the U.S. Government. The objective of Phase III, is for the small business to pursue commercialization objectives resulting from the Phase I/II R/R&D activities. The Small Business Act[3] defines commercialization as
  - "the process of developing products, processes, technologies, or services; and"
  - "the production and delivery (whether by the originating party or by others) of products, processes, technologies, or services for sale to or use by the Federal Government or commercial markets. "

  A significant advantage to Phase I/II award winners is that federal agencies may pursue sole source contracts to utilize technology developed through prior SBIR/STTR awards which automatically qualify as a Phase III activity.

[3] 15 U.S.C. § 638(e)(10).

The SBIR/STTR program needs to consider the success at attracting qualified small businesses to the program, the successful transitions between phases, and the impact of the technology in terms of commercialization.

d. Socially Disadvantaged and Underserved Populations

This particular report focuses on SBIR/STTR outreach to socially disadvantaged, and underserved areas.

Agencies have been increasing outreach efforts in the most recent years, but there is still much more work to be done. There is a need to improve outreach to women, minorities, and the 27 underserved states and territories as previously identified from the SBIR/STTR annual reports and independently by National Academy of Sciences (NAS) studies[4,5] to increase SBIR/STTR participation. SBA and the partnering 11 federal agencies have begun collecting data on how big the fissures are between these groups and the number of SBIR and STTR awards made each year.

Data reported shows the following concerning awards to these underserved populations:

- Data on women and minority program participation
    - From 2010-2013, awards to women owned firms in both the SBIR and STTR program have increased from 7.43% to 9.22%.
    - From 2010-2013, awards to minority owned firms have grown from 3.39% to 4.57% with a slight dip in between those years.
    - The peak years for awards were 1995 at 9.59% for women and 1994 for minorities with 9.14%, which we believe was partially due to a larger budget during that time period.
- Data on underserved areas
    - Awards have mostly been made to small businesses located in California, Massachusetts, New York and the Washington, DC metro area (combining Virginia, Maryland and the District of Columbia)
    - Those areas have won over 50% of all SBIR/STTR awards

[4] http://www.nap.edu/catalog.php?record_id=11989 (An Assessment of the SBIR Program)

[5] http://sites.nationalacademies.org/PGA/step/sbir/

We note that there is anecdotal evidence from various SBIR/STTR agency program managers regarding potential under-reporting by awardees that are socially disadvantaged or small businesses located in underserved communities. There is a newly created working group, which is part of the President's Lab to Market Commercialization Agenda under the Fueling Small Business Innovation Research section that is reviewing this issue as well as researching all the issues and recommendations set forth in this report. The group will also determine how the program can improve overall participation rates among stakeholders and where and how gaps can be minimized. The working group is currently reviewing the best practices and metrics to provide feedback and a path forward from the technology development perspective for the OSTP and SBA co-led IPC meetings.

3. Recommendations/Next Steps

In summary SBA and the 11 federal agencies will continue to work in tandem and collaboratively through the Outreach & Communications working group within the interagency policy committee team for the President's Lab to Market Commercialization agenda around Fueling Small Business Innovation. Below are key recommendations and next steps, many of which are already occurring and will continue to grow and build off as we move forward.

a. Unify Outreach Efforts

SBA along with the 11 federal agencies has taken the lead on creating an inter-agency outreach plan to provide for cost effectiveness and to ensure that efforts are not duplicated via the Outreach and Communications working group of the Fueling Small Business Innovation under the President's Lab to Market Commercialization agenda. However, all agencies will assess their current outreach structures and plans to identify gaps and create tactics to remove the missing parts, while goal aligning where practical under the SBIR/STTR banner.

b. Implementation of New Methods of Communication and Outreach Tools

The outreach and communications working group has put in place a game-plan to use both technology and face-face education and outreach efforts to targeted groups in order to increase their participation in the SBIR/STTR program. The use of various information technology driven social media platforms (Twitter, LinkedIn, Facebook, etc…) will help agencies leverage their resources and provide a farther reach to find new and diverse participants in the program. The program will tap government (federal, state and local), industry, and academic sources for new avenues of outreach.

Over the years, all of the 11 federal agencies involved with the SBIR/STTR program, have in concert and independently conducted numerous webinars promoting their various solicitations. Recently, SBA partnered with the National Council of Entrepreneurial Technology Transfer (NCET2) to put on a free 2 year webinar series[6] around the SBIR/STTR program with the specific purpose of increasing awareness and outreach to new innovation pipelines across the country. This webinar series (modeled after a similar focused series by NCET2 around IP-Law and SMEs[7]) is being conducted under the auspices of the White House Lab to Market Commercialization-Fueling Small Business Innovation interagency working group. A few examples of the webinar topics presented include updates on the Reauthorization Act, Commercialization, Understanding a specific Federal Agency's SBIR/STTR Program and Structure (e.g., DHS, DOE, DARPA, and NIH, etc…). We plan to continue this type of outreach to targeted groups of people. All of the material created via these webinar series is housed under the SBIR.gov website for anyone interested in the program to utilize and learn more.

Additionally the 11 federal agencies work in concert to help facilitate the execution of The National Innovation Summit and National SBIR/STTR conference each year which welcomes thousands to DC to learn more about the program and facilitate an open air of awareness and communication between the public and private stakeholders involved in the program.

---

6 http://center.ncet2.org/index.php?option=com_content&view=article&id=690
7 http://center.ncet2.org/index.php?option=com_content&view=article&id=445

c. Crafting the Right Narrative

Over the past year the 11 federal agencies, during the monthly SBIR/STTR program managers meetings and beyond, have been highlighting success stories for greater awareness internally and externally. With these success stories, we can better convey the mission and fruitful benefits of this program. From small business creation to workforce development to emerging industries that can trace its origin to the SBIR/STTR program. For instance it was recently found out from research conducted by the Institute for Defense Analysis-Science and Technology Policy Institute,that mainstream commercial 3D Printing can trace its seed financing origins in part to the SBIR/STTR program dating back to the early 1990's.[8] This type of information is very useful and critical in better understanding the benefits of this program and how each of the 11 federal agencies play crucial role in seeding next generation technology development and deployment.

d. Outreach to New Innovation Pipelines with the Underserved Communities

To ensure better awareness about the program, will require the right tool sets to help educate prospective applicants on how to participate in the SBIR/STTR programs across the 11 federal agencies. It is pertinent that accurate information from all of the stakeholders is gathered and disseminated. This is an issue being addressed by the working groups so that we can ensure a better managed and effective program that entices the appropriate participants across all 50 states. In particular the 27 states (AK, AR, DC, DE, HI, IA, ID, KS, KY, LA, ME, MO, MS, MT, ND, NE, NV, OK, PR, RI, SC, SD, TN, UT, VT, WV, WY) with the lowest success in the SBIR/STTR program will be a major focus of outreach awareness efforts to ensure that we are adequately tapping into the right innovation pipelines and encouraging the next generation of high-tech growth entrepreneurs. Key contacts have been identified from these 27 states in addition to all of the states with higher rates of success, with the aim to work closely with them on our outreach efforts. Additionally the 11 federal agencies are working on efforts to serve historically underserved communities through outreach efforts with a number of external stakeholder organizations. Approaching these various groups, identified below, will better

[8] https://www.ida.org/~/media/Corporate/Files/Publications/STPIPubs/ida-p-5091.ashx

help the SBIR/STTR program to reach the ideal applicants to participate and encourage buildout of next generation science and technology development.

The outreach to particular groups include:

- Historically Black Colleges and Universities (HBCU)
- National Society of Black Engineers
- Society of Hispanic Professional Engineers
- Society of Asian Scientists and Engineers
- American Indian Science and Engineering Society (AISES)
- Minority Business Development Agency (MBDA)
- US Black Chamber of Commerce
- US Hispanic Chamber of Commerce
- Society of Women Engineers;
- Association for Women in Science
- Minority Serving Institutions
- SBA District Offices
- Small business Development Centers
- Women's Business Centers
- SCORE chapters
- HubZones
- National Women's Business Council
- Universities for Women
- Society of Women Engineers
- Association for Women in Science
- Minority Serving Institutions, Societies, Associations, and Trade Organizations
- National Association for Equal Opportunity in Education
- State Science & Technology Institute
- National Conference of State Legislatures
- National Governors Association
- LGBT Technology Partnership
- Veterans-Related Service Groups

Within the resources available to them, agencies should work together in uniformity through the SBA to do outreach to targeted groups and underserved areas.

1) Hold the annual conference in an area that is underserved to encourage more companies in the underserved areas to participate in the program. Work with special interest chambers of commerce (e.g., Women, Hispanic, Black, LGBT etc.) will also encourage diverse participation in the program.
2) Build a mentorship program within SBIR so that experienced companies may recruit and mentor newer companies.
3) Create recruitment materials for various mediums that include foundational content such as SBIR and STTR 101, how-to guides for new entrants to the program, and success stories. These recruitment guides can be state specific and also highlight either woman owned firms or social disadvantaged owned firms.
4) Reach out to state economic development agencies to educate them on the SBIR and STTR programs so that they may tell their citizens to participate.
5) Work with universities in underserved states to educate them on SBIR and STTR.
6) Allocate administrative funds to specifically enable outreach for SBIR and STTR participation.
7) Hold regional conferences in areas that are underserved specifically to do outreach and increase applications by small business located in these areas.
8) Facilitate further encouragement and creation of student internship programs that encourage young people coming out of school to create their own companies and eventually apply to the SBIR program for initial R&D funding.
9) Continue to create and promote webinars that are content specific to reach out to socially disadvantaged small businesses and small businesses located in underserved communities.
10) Create target metrics, goals, and outcomes related to socially disadvantaged and underserved state participation.
11) Use proactive email marketing through the SBIR website to forward solicitations to companies that have not been involved in SBIR. This can be done using GovDelivery.
12) Utilize social media to tap the younger tech based sector.
13) Combine outreach efforts amongst agencies that cross over industries to strategically source outreach money.
14) Create and fund Accelerators and also educate ones that exist currently on the SBIR program.

15) Contact state and local based organizations and small business service providers in underrepresented areas to increase awareness among small businesses residing in these zones.
16) Upgrade website materials and content to recruit women, socially disadvantaged individuals, and small businesses from underserved areas.
17) Hold entrepreneurial boot camps for new applicants, which would be targeted towards socially disadvantaged owned small businesses and small businesses located in underserved communities.

In: Small Business Innovation Research ... ISBN: 978-1-63482-532-0
Editor: Thomas H. Chavez

Chapter 5

SBIR/STTR COMMERCIALIZATION*

SBIR/STTR Interagency Policy Committee

1. EXECUTIVE SUMMARY

Purpose. This report responds to the statutory requirement in the Small Business Innovation Research (SBIR) / Small Business Technology Transfer (STTR) Reauthorization Act of 2011 (Reauthorization Act), which was included in the 2012 Defense Authorization Act (Pub. L.112-81)[1]. The Reauthorization Act requires the Interagency Policy Committee (IPC) to review the issues around successful commercialization of SBIR/STTR funded technologies.

Key Issues. The key issues that need to be addressed include: improving compliance; reducing processing times; improving commercialization; and measuring the value of the SBIR / STTR programs via agreed upon benchmarks by the agencies from a technology commercialization perspective. This report focuses on the best practices that Federal agencies have implemented to increase commercialization results for SBIR and STTR awardees. This report also focuses on new ways to address the funding gaps that SBIR/STTR awardees often face after the receipt of a Phase II award but

* This is an edited, reformatted and augmented version of a report to Congress issued by the Small Business Innovation Research (SBIR) and Small Business Technology Transfer (STTR) Program Interagency Policy Committee, September 15, 2014.

[1] http://www.gpo.gov/fdsys/pkg/PLAW-112publ81/pdf/PLAW-112publ81.pdf

before commercialization. Many of these issues will overlap the IPC Standard Evaluation Framework and TechNet Public & Government Databases reports and thus should be read in conjunction with another to better understand these issues holistically.

Key Recommendations:

- The criteria shown in the diagram below should be assessed for feasibility and budgetary constraints. If acceptable, changes to the SBIR.gov database, overseen by SBA, should be made to collect and automate metrics in determining best practices for technology commercialization based on the subject matter of interest.

Outreach
- New Applicants
- Women and Disadvantaged
- Geographic
- Outreach Activity Assessment

Compliancy & Efficiency
- Award levels
- Data requirements
- Processing Times
- Fraud, Waste & Abuse

Commercialization & Other Economic
- Commercialization
    - Phase I to Phase II Conversion
    - Initial
    - Full
- Return on Investment
- Jobs Created/Sustained

- To better gauge and evaluate commercialization efficacies across the SBIR/STTR programs and facilitate continued maintenance of a centralized database that allows for continuous and up-to-date information to flow freely between SBA and the Federal agencies that administer SBIR/STTR awards. This will enable an in-depth understanding of data across the 11 Federal agencies on how better commercialization of SBIR/STTR funded projects might occur based on prior success and understanding of how that success came about.
- Focused effort across the SBIR/STTR programs to increase success of Tech-to-Market initiatives. Federal agencies will continue to share best practices to improve commercialization results for SBIR/STTR awardees.[2] This includes review and understanding around leveraging

[2] DOD supports collaboration whenever practical, however as stated previously the agency does not support the use of interagency/joint funding outside the DOD community.

Commercialization Assistance Programs used by some of the larger and smaller agencies.
- Ensure better usage and promotion of various technology accelerator and incubator programs that help with technology-risk mitigation and foster greater success for high tech startup growth.

2. INTRODUCTION

The Reauthorization Act, a division of the National Defense Authorization Act for Fiscal Year 2012, created the SBIR/STTR Interagency Policy Committee (IPC). The IPC is co-chaired by the White House Office of Science and Technology Policy (OSTP) and the Small Business Administration (SBA) and includes representatives from Federal agencies that participate in the SBIR or STTR programs. As part of the Reauthorization Act, the IPC is required to review certain issues and make policy recommendations to Congress on ways to improve program effectiveness and efficiency.

a. Background

Congress established the SBIR program in 1982 and the STTR program ten years later to assist small business concerns (SBCs) in obtaining Federal research and development (R&D) funds to build a strong economy and support technological innovation as discussed below:

- *SBIR:* Requires Federal agencies with extramural R/R&D budgets over $100 million to set aside a percentage of their annual extramural R/R&D budget for small businesses. This percentage was 2.5% prior to FY 2012, and increased to 2.6% in FY 2012 and will continue to increase by 0.1% each year until it reaches a base requirement of 3.2% in FY 2017.
- *STTR:* Modeled after the SBIR program, STTR requires Federal agencies with extramural budgets exceeding $1 billion to set aside a percentage of their annual extramural R&D budget for SBCs that work in cooperation with universities, federally funded research and development centers, and other non-profit scientific and educational institutions. This percentage was 0.3% in FYs 2004-2011, and

legislation increased this minimum to 0.35% for FYs 2012 and 2013, with continued increases through 2016. The goal is to facilitate transfer of technology and research from these institutions to commercial use and encourage innovation.

b. Program Objectives

The Small Business Act indicates that "assistance" provided by the SBIR and STTR programs "be given to small-business concerns to enable them to undertake and obtain the benefits of research and development in order *to maintain and strengthen the competitive free enterprise system and the national economy.*"[3]

The SBIR Policy Directive identifies the following primary objectives:

- Stimulate technological innovation;
- Meet Federal R&D needs while strengthening the role of small businesses;
- Foster and encourage participation in innovation and entrepreneurship by socially and economically disadvantaged persons; and,
- Increase private-sector commercialization of innovations derived from Federal research and development funding.

In addition, the Reauthorization Act added several new initiatives including:

- Permitting agencies to direct some SBIR funds to firms that are owned by multiple venture capital operating companies (VCOCs), hedge funds, or private equity firms;
- Allowing agencies to provide one additional Phase II award to small businesses to extend a study;
- Commercialization readiness program at DOD and pilot programs at civilian agencies;
- Reducing processing times;
- Pilot program to allow agencies to use 3% of program budgets for administration and oversight;

[3] 15 U.S.C. § 638(a).

- Reducing vulnerability of fraud,waste, and abuse (FWA);
- Provisions for improved program evaluation; and,
- Other initiatives to increase commercialization and outreach.

In general, the SBIR/STTR reauthorization legislation underscored the need for improved commercialization, outreach, and program evaluation.In December 2013 the lead SBIR/STTR Program Managers in conjunction with White House Office of Science & Technology Policy (OSTP) facilitated the creation of five sub-working groups that fall under the "Fueling Small Business Innovation" component of the President's Lab to Market Commercialization Agenda. These five groups are:

1. Outreach & Communications
2. Commercialization Pathways
3. Awards Efficiency & Efficacy
4. Databases & Interagency Exchange of Information
5. Asset Mapping

Going forward these five groups are tasked with various short-term and long-term projects that fall within scope of various facets covered in the President's Lab to Market Commercialization Agenda. They will seek to identify issues, challenges, and provide further recommendations for consideration amongst the various SBIR/STTR program managers as well for SBA and OSTP consideration as it relates to the SBIR/STTR program.

c. SBIR/STTR Process

To bring technology from ideas to commercialization, both programs utilize a three phase approach:

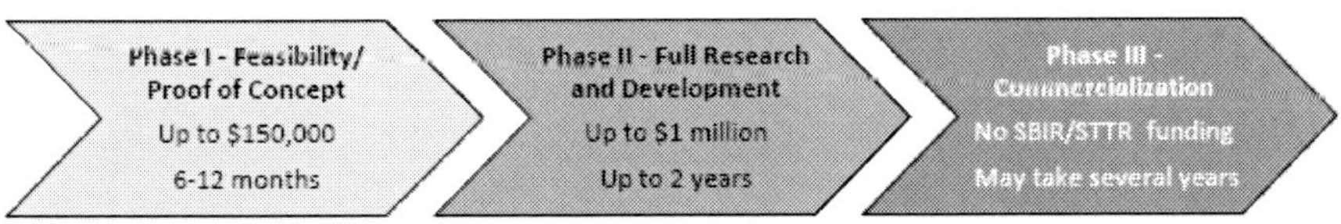

- *Phase I – Feasibility/Proof of Concept.* Using a competitive process, Federal agencies award up to $150,000 to a small business to perform research/research and development (R/R&D) for up to 6-12 months on a specific topic in order to establish its technical merit, feasibility, and commercial potential. During this phase, Federal agencies assess both the performance of the small business and the potential of the technology prior to providing further Federal support in Phase II.

- *Phase II – Full Research and Development.* Based on the results achieved in Phase I, Federal agencies will decide whether to continue R/R&D efforts into Phase II based on the scientific, technical, and commercial merit and feasibility of the idea. If the Federal agency decides to continue into Phase II, they will award up to $1 million to the small business to continue R/R&D efforts for up to 2 years.
- *Phase III - Commercialization.* No specific SBIR funding is associated with Phase III, however, some agencies may include follow-on non-SBIR funded R&D or production contracts for products or services intended for use by the U.S. Government. The objective of Phase III, is for the small business to pursue commercialization objectives resulting from the Phase I/II R/R&D activities. The Small Business Act[4] defines commercialization as:
  - the process of developing products, processes, technologies, or
  - the production and delivery (whether by the originating party or by others) of products, processes, technologies, or services for sale to or use by the Federal Government or commercial markets.

  A significant advantage to Phase I/II award winners is that Federal agencies may pursue sole source contracts to utilize technology developed through prior SBIR/STTR awards which automatically qualify as a Phase III activity.

The IPC's reports consider whether Federal agencies participating in the SBIR/STTR programs have successfully attracted qualified small business concerns to participate in the program, whether SBIR/STTR award recipients experience successful transitions between the phases of the programs, and the impact of the technology developed through SBIR/STTR awards as evidence of commercialization success.

[4] 15 U.S.C. § 638(e)(10).

3. SBIR Commercialization and Government Agencies

a. Overview

Commercialization as defined by the Small Business Act is: The process of developing products, processes, technologies, or services and the production and delivery (whether by the originating party or others) of the products, processes, technologies, or services for sale to or use by the Federal government or commercial markets.[5]

Increasing private sector commercialization of innovations derived from SBIR/STTR funds is one of the key pillars of the program.

- Commercialization Database[6]:Under the Reauthorization Act, SBA is required to build a commercialization database (further discussed in the IPC TechNet Public & Government Databases report) to track the performance of each SBIR/STTR awardee and conduct portfolio analysis. Information procured includes: number of awards/contracts/grants procured, patent related applications filed as a result of funding, number of employees, structure of ownership, etc.
- Having this information will be key to evaluating the performance and success of the program from a commercialization efficacy prospective. Within this mission, applicant firms must now register at a Company Registry Database on SBIR.gov, which will be centralized to lower the agency's reporting burden and give cross-agency visibility into firms' performance history. However, unless stated otherwise, the information contained in the Company Registry Database, the Application Information Database, and the Commercialization Database are solely available to authorized officials, with the approval of SBA. *Working with agencies, the Small Business Administration should explore options for providing resources for building and maintaining databases and other systems to support SBIR/STTR program requirements. Options for resources could include a direct request for appropriations to perform these functions to be managed by SBA, use of an interagency funding*

[5] 15 U.S.C. § 638(e)(10).

[6] DOD supports collaboration whenever practical, however as stated previously the agency does not support the use of interagency/joint funding outside the DOD community.

agreement between SBIR/STTR agencies and SBA, or other funding mechanisms.

- Annual Report: When submitting their report to SBA annually, agencies must include an accounting of Phase I awards made to SBCs that have received more than 15 Phase II awards from all agencies in the preceding 5 fiscal years (In general all Phase I awards are reported, but this particular metric is crucial for the commercialization benchmark perspective). Each agency must report: the name of the awardee; Phase I funding agreement number and date of award; Phase I topic or subtopic title; amount and date of previous Phase II funding; and commercialization status for each prior Phase II award. *Each agency must submit a report to SBA on those SBCs that submitted an application and were found to not meet the agency's benchmarks with respect to progress towards commercialization. This report must include the name and employer identification number of the SBC, the closing date of the solicitation to which it proposed, and the agency that issued the solicitation.*
- Agency Assistance: Participating agencies within the SBIR/STTR program will soon be disseminating datasets that relate to the central work of their agency. In many cases, these datasets on their own tell only a limited story, however, when combined with data from other agencies charged with assisting the small business community with job creation and private sector investment success, the potential for a powerful assessment tool is introduced. *The SBIR/STTR Commercialization Database should be able to mine as well display,as appropriate, individual agency data and actions in order to better promote the creation of American jobs through greater participation, collaboration and transparency.*

b. Commercialization Assistance

- A number of SBIR/STTR agencies have been very successful at helping their awardees commercialize. However, there is no streamlined way for agencies to share lessons learned. SBA should lead in the sharing of best practices among SBIR/STTR agencies and help lay out the appropriate steps to achieve that leverage as well as proposals for new initiatives to address funding gaps that business concerns face after Phase II, but before commercialization.

SBIR/STTR participants must keep in mind that per reauthorization, a Federal agency may provide to the recipient of a Phase I SBIR or STTR award, an amount equal to but not more than $5,000 per year in order to purchase the services of a vendor that would assist with commercialization efforts. The same applies to the recipient of a Phase II SBIR or STTR award. In carrying out these efforts, each Federal agency shall provide the allowable amounts to a recipient that meets the eligibility requirements, even if the recipient requests to seek technical assistance from an individual or entity other than the vendor selected by the Federal agency.

c. Commercialization Readiness Program DoD

- General: The Department of Defense is authorized to create and administer a "Commercialization Readiness Program" to accelerate the transition of technologies, products, and services developed under the SBIR Program to Phase III, including the acquisition process. *The authority to create this Commercialization Readiness Program does not eliminate or replace any other SBIR or STTR program that enhances the insertion or transition of SBIR or STTR technologies.*
- Accelerated transition of research programs to acquisition process: DoD must identify research programs of the SBIR Program that have the potential for rapid transitioning to Phase III and into the acquisition process and certify in writing that the successful transition of the program to Phase III and into the acquisition process is expected to meet high priority requirements.

d. Commercialization Readiness Pilot Program for Civilian Agencies

- General: The Commercialization Readiness Pilot Program permits the head of any Federal agency participating in the SBIR Program (except DoD) to allocate not more than 10% of its funds allocated to the SBIR Program in two modes:

    - For follow-on awards to small businesses for technology development, testing, evaluation, and commercialization assistance for SBIR or STTR Phase II technologies; or
    - For awards to small businesses to support the progress of R/R&D and commercialization conducted under the SBIR or STTR programs to Phase III.
- Application to SBA: Before establishing this pilot program, the agency must submit a written application to SBA not later than 90 days before the first day of the fiscal year in which the pilot program is to be established. The written application must set forth a compelling reason that additional investment in SBIR or STTR technologies is necessary, including unusually high regulatory, systems integration, or other costs relating to development or manufacturing of identifiable, highly promising small business technologies or a class of such technologies expected to substantially advance the mission of the agency.
- SBA's Determination: SBA must make its determination regarding an application to the pilot program, not later than 30 days before the first day of the fiscal year for which the application is submitted. SBA must also publish its determination in the Federal Register and make a copy of the determination and any related materials available to the Committee on Small Business and Entrepreneurship of the Senate and the Committee on Small Business and the Committee on Science, Space, and Technology of the House of Representatives.

e. Methodology

- SBA is currently building the SBIR/STTR Commercialization Database with routine User Acceptance Testing (UATs) provided by the agencies in order to produce an outcome that is unanimously supported due to continuous implementation practical agency feedback. A primary objective throughout the entire process has been a seamless transition due to careful analysis of individual agency database questions, while adhering to congressional requirements for displaying program success via performance metrics (sales, revenue, job created, patents, etc.). The three-phase process of build-up and implementation is slated to be complete by late 2014.

f. SBIR/STTR Commercialization Best Practices

- Since the program's inception, a wealth of information and institutional knowledge has been procured about how to best successfully reach effective commercialization. As one would suspect, the notion and timeline readiness for technology market maturation varies differently by the sector for a variety of reasons from "financially intensive readiness" to an appropriate "technology support system" to fuel growth and development of a particularly advanced piece of hardware or software.
- Commercialization Assistance Programs (CAPs) are generally third party subcontractors that aid in the mentoring and initial scaling of small businesses that are seeking mentoring, advising, and network building as they reach proof of concept stage for prototype development (Phase I) or endeavor in scale up commercialization opportunities for market expansion from an initial customer base (Phase II). Typical programmatic endeavors that result from CAPs (like Dawnbreaker, Larta, Foresight, etc.) that aid in development include webinars, workshops, annual conferences, networking opportunities, etc.

Additionally, there are various types of in between "Post Phase Programs" that are utilized to help derive better chances of successful commercialization in bridging the gap over the "Valley of Death". These include but are not limited to:

- Post Phase II Bridge funding (e.g. Phase II-B/II-E/II-X)
- Phase III Awards (e.g., Federal acquisition of SBIR products – DoD (especially Navy), NASA, HHS (through Biomedical Advanced R&D Authority), DHS, etc.)
- Commercialization Readiness Programs

4. Conclusion & Recommendations

a. Proposals for New Initiatives

- Leveraging partnerships with high growth stakeholders: There are a multitude of organizations that provide resources to high growth entrepreneurs. SBA will take the lead in a "train the trainers" model and reach out to these trade associations, high growth groups, and universities to train them about the SBIR/STTR program. Through these relationships, SBA can:
  - Conduct outreach activities to these stakeholder's members (i.e.: webinars, educational materials, etc.)
  - Connect these stakeholders to specific agencies for additional outreach activities by sector or by region.
  - Attend stakeholder conferences where Program Managers can rotate between those based on region and sector. This way, Program Managers are able to target their travel while doing outreach to individuals that are most likely to apply to their solicitations and commercialize.
  - Explore how NSF's I-Corp model might be applicable and/or deployable amongst other agencies (HHS-NIH recently implemented usage of the program; DHS will implement its own I-Corp model in FY 2015).

Accelerators, Incubators, and Clusters: Continuing along the path of resources to small businesses, there are a host of available avenues that small businesses can pursue either directly at SBA or through assistance from the agency. These resources can play the role of intermediary or mentor to provide small business with the tools, funding, and guidance required to achieve successful commercialization. *SBA will take the lead to work with their partners at the Office of Entrepreneurial Development (OED) and the global accelerator/incubator network to schedule innovation summits, conferences, and demo day allowing for introduction and substantial interaction between the various entities.*

b. Key Recommendations

- Policy Recommendation 1: Leveraging CAP programs at larger agencies: The SBIR/STTR program should analyze the data on the current commercialization assistance programs.SBA should look into smaller SBIR programs that would like to take advantage of the services offered by larger agencies, but cannot get contractor interest because of their small size. If an interest is indicated, SBA should explore funding mechanisms whereby smaller agencies can collaboratively work with larger agencies to obtain those services.
- Policy Recommendation 2: SBA and the SBIR/STTR participating agencies should host meetings with CAP providers and small businesses to optimize offerings from these programs For example,the feasibility of hosting a session during a National SBIR conference whereby the contractors provide presentations to SBA and all agencies on the strengths and weaknesses of these programs based on their interactions with small businesses should be investigated.Furthermore, several small businesses should be tapped to share their experiences, positive and negative, with these commercialization services. (Session could be closed or open to the public.)
- Policy Recommendation 3: By the end of 2016, agencies should collect feedback on small businesses that utilized their own commercialization assistance providers and analyze the information to make future recommendations for this statutory provision.

In: Small Business Innovation Research ... ISBN: 978-1-63482-532-0
Editor: Thomas H. Chavez

Chapter 6

SBIR/STTR STANDARD EVALUATION FRAMEWORK*

SBIR/STTR Interagency Policy Committee

1. EXECUTIVE SUMMARY

Purpose. The Small Business Innovation Research (SBIR) / Small Business Technology Transfer (STTR) Reauthorization Act of 2011 (Reauthorization Act), which was included in the 2012 Defense Authorization Act (P.L. 112-81) signed into law on December 31, 2011, charges the Interagency Policy Committee (IPC) with, "Developing and incorporating a standard evaluation framework to enable systematic assessment of SBIR and STTR, including through improved tracking of awards and outcomes and development of performance measures for the SBIR program and STTR program of each Federal agency."

Key Issues. The key issues to be addressed by the evaluation framework include: measuring the level of outreach to small businesses underrepresented in the SBIR/STTR programs; determining Federal agency compliance with statutory and policy directive guidance; determining whether SBIR/STTR award processing timelines were reduced; improving commercialization; and measuring the overall value of the SBIR/STTR programs.

---

* This is an edited, reformatted and augmented version of a report to Congress issued by the Small Business Innovation Research (SBIR) and Small Business Technology Transfer (STTR) Program Interagency Policy Committee, September 15, 2014.

Key Recommendations.

- It is recommended that the IPC adopt a six-step evaluation framework as follows:
  1) Engage Stakeholders – Completed with the formation of the IPC;
  2) Describe the Program – Completed with the Reauthorization Act and Policy Directive;
  3) Focus the Design – Completed by adopting the recommendations in this report;
  4) Gather the Data – Already initiated;
  5) Draw Conclusions – SBA and the five SBIR/STTR agencies with an SBIR Program budget of more than $50,000,000 in fiscal year 1999, have entered into an agreement with the National Academy of Sciences (NAS) to conduct an independent study of the selected elements of the two programs.Additionally the agencies through the IPC Fueling Small Business Innovation Working Group, will conduct their own internal series of mini evaluation projects regarding pertinent areas of importance for the SBIR/STTR program in better understanding the efficiency and efficacies of the program; and,
  6) Present Findings – These will be presented for review and understanding wherein applicable.
- The criteria identified in Appendix A (shown in the diagram below) should be assessed for feasibility and budgetary constraints. If acceptable, changes to the database should be made to collect and automate metrics.

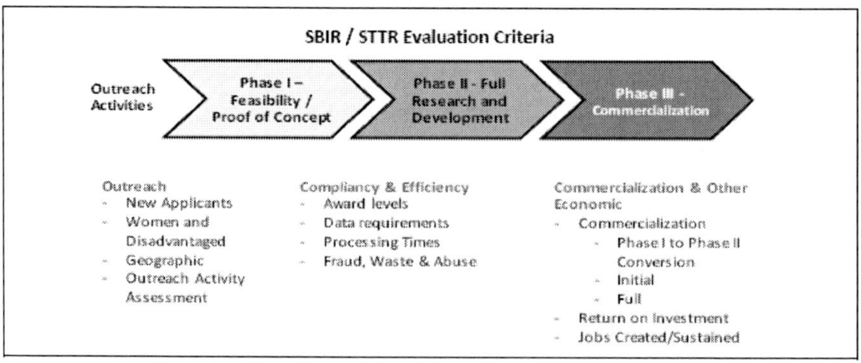

2. INTRODUCTION

The Small Business Innovation Research (SBIR) / Small Business Technology Transfer (STTR) Reauthorization Act of 2011 (Reauthorization Act), a division of the National Defense Authorization Act for Fiscal Year 2012(P.L. 112-81)[1] signed into law on December 31, 2011, created the SBIR/STTR Interagency Policy Committee (IPC). The IPC is co-chaired by the White House Office of Science and Technology Policy (OSTP) and the Small Business Administration (SBA) and includes representatives from Federal agencies that participate in the SBIR or STTR programs. As part of the Reauthorization Act, the IPC is required to review certain issues and make policy recommendations to Congress on ways to improve program effectiveness and efficiency.

This report reviews the issues regarding the development and implementation of a standard evaluation framework to assess the SBIR and STTR programs.

a. Background

Congress established the SBIR program in 1982 and the STTR program ten years later to assist small business concerns (SBCs) in obtaining Federal research and development (R&D) funds to build a strong economy and support technological innovation as discussed below:

- SBIR: The SBIR program requiresFederal agencies with extramural Federal research or research and development (R/R&D) budgets over $100 million to set aside a percentage of their annual extramural R/R&D budget for SBIR awards to small businesses. Prior to Fiscal Year (FY) 2012, agencies that met the budget threshold, were required to spend at least 2.5% of their extramural R/R&D budget on SBIR awards. This percentage increased to 2.6% in FY 2012 and will continue to increase by 0.1% each year until it reaches a base requirement of 3.2% in FY 2017.
- STTR: Modeled after the SBIR program, the STTR program requires Federal agencies with extramural R/R&D budgets exceeding $1 billion to set aside a percentage of their annual extramural R/R&D

[1] http://www.gpo.gov/fdsys/pkg/PLAW-112publ81/pdf/PLAW-112publ81.pdf.

budget for small business concerns (SBCs) that work in cooperation with universities, Federally funded research and development centers, and other non-profit scientific and educational institutions. This percentage was 0.3% in FYs 2004-2011, and legislation increased this minimum to 0.35% for FYs 2012 and 2013, with continued increases through 2016. The goal is to facilitatethe transfer of technology and research from these institutions to commercial uses and to encourage innovation.

b. Program Objectives

The Small Business Act indicates that "assistance"provided by the SBIR and STTR programs "be given to small-business concerns to enable them to undertake and to obtain the benefits of research and developmentin order *to maintain and strengthen the competitive free enterprise system and the national economy*."[2]

The SBIR/STTR Policy Directives identify the following primary objectives:

- Stimulate technological innovation;
- Meet Federal R&D needs;
- Foster and encourage participation in innovation and entrepreneurship by socially and economically disadvantaged persons; and,
- Increase private-sector commercialization of innovations derived from Federal research and development funding.

In addition, the Reauthorization Act added several new initiatives including:

- Permitting agencies to direct some SBIR funds to firms that are owned by multiple venture capital operating companies (VCOCs), hedge funds, and/or private equity firms;
- Allowing agencies to provide one additional Phase II award to small businesses to extend a study;
- The commercialization readiness program at the U.S. Department of Defense (DoD) and pilot programs at civilian agencies;

[2] 15 U.S.C. § 638(a).

- Reducing processing times between solicitation request closing dates and final agency decision dates of awardee selection;
- Pilot program to allow agencies to use 3% of program budgets for administration and oversight;
- Reducing the vulnerability of fraud, waste, and abuse (FWA);
- Provisions for improved program evaluation; and,
- Other initiatives to increase commercialization and outreach.

In general, the SBIR/STTR reauthorization legislation underscored the need for improved commercialization, outreach, and program evaluation.In December 2013 the lead SBIR/STTR Program Managers in conjunction with White House Office of Science & Technology Policy (OSTP) facilitated the creation of five sub-working groups that fall under the "Fueling Small Business Innovation" component of the President's Lab to Market Commercialization Agenda. These five groups are:

1) Outreach & Communications
2) Commercialization Pathways
3) Awards Efficiency & Efficacy
4) Databases & Interagency Exchange of Information
5) Asset Mapping

Going forward these five groups are tasked with various short-term and long-term projects that fall within scope of various facets covered in the President's Lab to Market Commercialization Agenda. They will seek to identify issues, challenges, and provide further recommendations for consideration amongst the various SBIR/STTR program managers as well for SBA and OSTP consideration as it relates to the SBIR/STTR program.

c. SBIR/STTR Process

To bring technology from ideas to commercialization, both programs utilize a three phase approach:

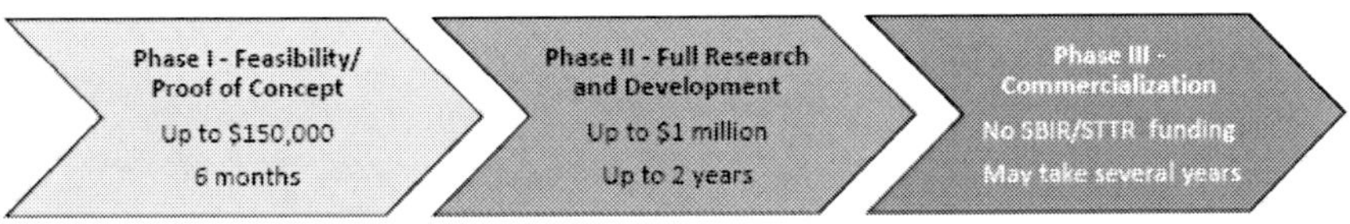

- Phase I – Feasibility/Proof of Concept. Using a competitive process, federal agencies award up to $150,000 to a small business to perform research/research and development (R/R&D) for up to 6-12 months on a specific topic in order to establishits technical merit, feasibility, and commercial potential. During this phase, federal agenciesassessboth the performance of the small businessand the potential of the technology prior to providing further Federal support in Phase II.
- Phase II – Full Research and Development.Based on the results achieved in Phase I, federal agencies will decide whether to continue R/R&D efforts into Phase II based on the scientific, technical, and commercial merit and feasibility of the idea. If the federal agency decides to continue into Phase II, they will award up to $1 million to the small business to continue R/R&D efforts for up to 2 years.
- Phase III - Commercialization. No specific SBIR funding is associated with Phase III. The objective of Phase III is for the small business to pursue commercialization objectives resulting from the Phase I/II activities. The Small Business Act3 defines commercialization as:
  - the process of developing products, processes, technologies, or
    - the production and delivery (whether by the originating party or by others) of products, processes, technologies, or services for sale to or use by the Federal Government or commercial markets. A significant advantage to Phase I/II award winners is that Federal agencies may pursue sole source contracts to utilize technology developed through prior SBIR/STTR awards which automatically qualify as a Phase III activity.

The SBIR/STTR program evaluation framework needs to consider and/or look into the success at attracting additional qualified small businesses to compete for agency awards within the programs, the successful transitions between phases of all small businesses receiving SBIR/STTR awards, and the impact of the technology in terms of commercialization.

[3] 15 U.S.C. § 638(e)(10).

3 Discussion

a. SBIR/STTR Program Evaluation Framework

A program evaluation framework summarizes and organizes the steps and standards for effective program evaluation. The chart below shows the steps in developing and evaluating a program.

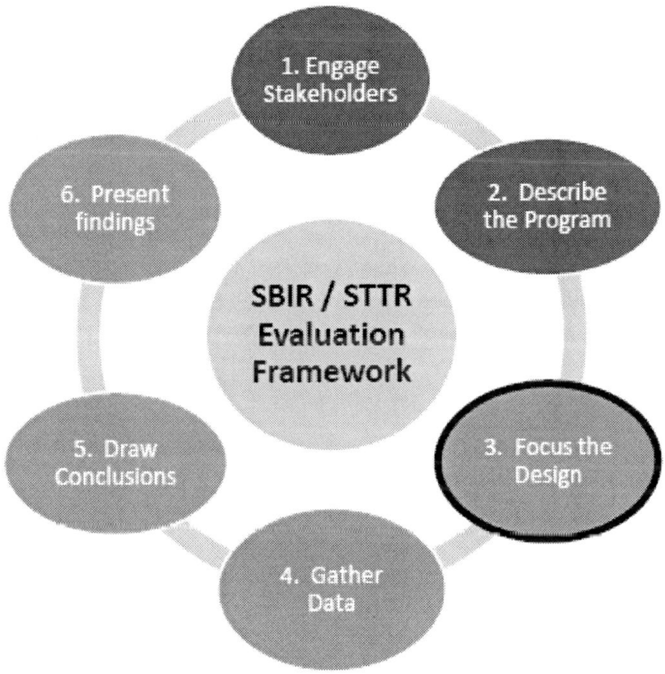

As shown, the process is cyclical where the evaluation criteria are developed based on input by the stakeholders and assessing program objectives and processes. After focusing the design by establishing evaluation criteria, data is gathered and evaluated. Based on the conclusions drawn and findings that are presented, the IPC may make recommendations to improve the effectiveness and efficiency of the programs.

The U.S. Government Accountability Office (GAO) has conducted several assessments of the programs' effectiveness. In addition, in 2008, the National Academy of Sciences (NAS) provided a report to Congress assessing the SBIR/STTR programs. In general, the GAO and NAS studies found the programs to be effective in achieving the program objectives (see 2b in this report), but cited the need to improve data collection with an emphasis on commercialization and outreach to women-owned and socially or economically disadvantaged business owners. Many of the recommendations cited in the NAS study were incorporated into the Reauthorization Act of 2011.

With this most recent reauthorization, the SBIR/STTR evaluation framework must be reassessed to meet modified objectives. The goal is to provide the analysis necessary to improve the efficiency and effectiveness of the programs and support policy making efforts wherein possible amongst the collective efforts of the federal agencies involved in the administering of the SBIR/STTR program as it relates to each agency's mission directive in applicable conjunction with the program itself. Each of the above steps is discussed below with this goal in mind.

- *Step 1 – Engage the Stakeholders.* With the formation of the IPC, stakeholder representatives of the SBIR/STTR programs and the general American innovation ecosystem are engaged.
- *Step 2 – Describe the Program.* The Reauthorization Act and subsequent updated SBIR/STTR Policy Directives enacted and implemented several changes to the programs. Section 2b in this report summarizes the programs, their objectives, and the major statutory changes to the programs.
- *Step 3 – Focus the Design.* In this step, stakeholders focus the design of thte evaluation framework by establishing evaluation criteria which assess the issues of greatest concern considering both time and budgetary constraints. This report focuses on Step 3.
- *Step 4 – Gather the Data.* The data needed to assess the program will be collected via a common SBIR database. One of the IPC reports describes the commercialization database that will be implemented as required by Subtitle C of the Reauthorization Act.
- *Step 5 – Draw Conclusions.* Section 5137 of the Reauthorization Act requires SBA to enter into an agreement with NAS to continue their independent study of the SBIR and STTR programs 4 years after enactment (by December 31, 2015) and every 4 years thereafter.
- *Step 6 – Present Findings.* The NAS Study will be presented to the agencies who fund the study for review and technical comment. The IPC may also review the study. The study and any recommendations will be provided to Congress to support policy making efforts.

b. SBIR/STTR Evaluation Metrics and Standards

The NAS Study acknowledged the difficulty in assessing the value of the SBIR/STTR program, given the different missions of each of the program's participating agencies. As a result, the Reauthorization Act required each participating agency to develop metrics to evaluate the effectiveness and the benefit to the people of the United States of the SBIR program and the STTR program of the Federal agency that:

(A) are science-based and statistically driven;
(B) reflect the mission of the Federal agency; and
(C) include factors relating to the economic impact of the programs.

The participating agencies agreed that a standard set of metrics are more appropriate in order to simplify the evaluation process. By examining the SBIR/STTR process and objectives, evaluation criteria can be categorized as shown in the following chart.

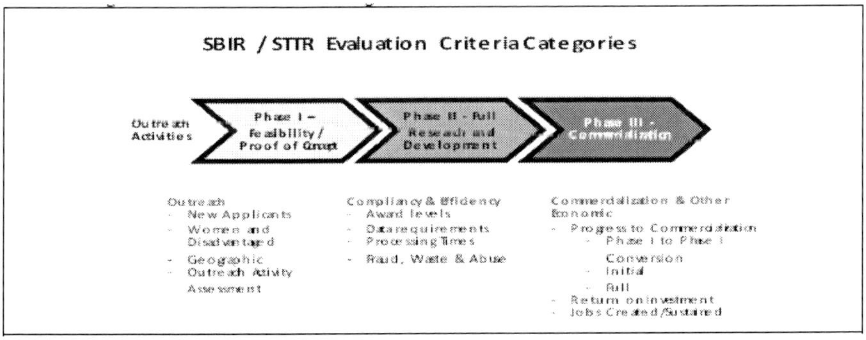

Each of these categories is described below:

1) *Outreach*– This category assesses both general and targeted outreach of the SBIR/STTR program. All agencies are better served by increasing the number of new qualified applicants to the program as a larger pool of qualified applicants would likely yield better returns for the agencies from a successful technology transfer, commercialization, and/or scale up in product deployment and growth opportunity. However, Congress has specifically identified the need to

increase outreach to entrepreneurs that have been historically underserved and socially disadvantaged. This includes businesses owned by women or socially and economically disadvantaged groups and reaching out to geographic areas where small business concerns are located but historically have not received many SBIR/STTR awards. In addition, some of the new Reauthorization Act initiatives are focused specifically on improved outreach. Assessing the effectiveness of these initiatives is also important.

2) *Compliance and Efficiency*– This category addresses participating agency compliance with the SBIR/STTR statutory requirements. This includes providing funding for SBIR and STTR awards that satisfies the minimum levels identified by statute, including the proper data requirements in SBIR and STTR awards, processing reduction improvements, and reduction in FWA.
3) *Commercialization & Other Economic* – Ultimately, SBIR/STTR program success depends on the commercialization of the technology created and supported through the SBIR and STTR awards into products and services available for purchase by the Government, private sector, and the public. Congress has concerns that there may be businesses that receive multiple awards and never achieve commercialization. Section 5165 of the Reauthorization Act requires the Administration to establish a minimum commercialization performance standard for continued receipt of Phase I awards. SBA published this commercialization requirement in the Federal Register, 78 Fed. Reg. 48537 (Aug. 8, 2013). The purpose of this requirement is to establish a certain level of commercialization success in order for a small business to continue to be eligible for the receipt of new Phase I awards. It may take several years to achieve a significant amount of sales and investment and therefore a more comprehensive measure of commercialization should be identified to evaluate the program's success.

Although SBIR/STTR commercialization success has historically been measured as the percentage of Phase II awards that achieve some level of commercialization, data should be collected so that the potential economic and technology development returns on SBIR & STTR award dollars can be measured. Job data should also be collected to understand the impact from these Federal programs.

Appendix A identifies proposed metrics/standards for each of the above categories. These metrics should be assessed for feasibility and budgetary constraints. In addition, the IPC should review metrics on a periodic basis for continued improvement.

4. RECOMMENDATIONS / NEXT STEPS

In summary, we recommend the following:

- It is recommended that the IPC adopt the six-step evaluation framework described in section 3a which includes: 1) Engage Stakeholders; 2) Describe the Program; 3) Focus the Design; 4) Gather the Data; 5) Draw Conclusions; and 6) Present Findings.
- It is recommended that further investigation be performed to assess the feasibility of the measurements identified in Appendix A (based on information gathered by NAS independent study and concurrently the IPC's Fueling Small Business Innovation working groups). If the needed data is unavailable or the measurement is too difficult or costly to generate, it may need to be removed.
- It is recommended that, after assessing feasibility, the SBA revise its commercialization database as necessary to collect the necessary data, automatically generate these metrics, and make the metrics available on its website at www.SBIR.gov.
- It is recommended that SBA continually seek advice and feedback from outside experts as to how to improve this evaluation framework.
- It is recommended that the IPC reassess this framework and evaluation criteria on a periodic basis to make changes as needed via the Fueling Small Business Innovation working groups.

APPENDIX A – PROPOSED EVALUATION CRITERIA

This appendix outlines the proposed evaluation criteria for the SBIR/STTR Program for each of the categories described in Section 3b.

a. Outreach

Assessment Questions / Rationale	Potential Proposed Metrics
1. ***New Applicants:***	
• Is the SBIR/STTR program attracting new applicants to the program? A new applicant is defined as a business that has not previously applied to the SBIR/STTR program.	• Number and percent of Phase I applications by new applicants
• Is the program increasing the number of awards to first-time award winners (businesses that have never received an award)?	• Number and percent of Phase I awards and award dollars to first-time award winners
• How many times do first-time award winners apply for an award before they receive an award?	• Number of times first-time award winners apply for Phase I award before receiving an award
• What is the cost to acquire a new applicant? (Although the cost to acquire a new applicant is difficult to assess, by looking at outreach activity costs in relation to new applicants or increase of new applicants may provide some indication of cost. If new applicants identify where they heard about the program, this may provide further insight.)	• Outreach activity costs to new applicants or increase of new applicants
• Are awards going to the same firms? Rationale: Similar to any business, understanding new and repeat "customers" is important. It helps assess the impact of educational and outreach activities. Also, one concern is that previous award winners are the primary recipients of SBIR/STTR awards. This may be because the pool of qualified applicants is not large enough. A larger pool of qualified new applicants will likely lead to improved program results. It is also important to understand the prevalence of multi-award winners.	• Pie Chart indicating the number of small businesses receiving Phase I awards by the number of Phase I awards received for a five year period, as shown below. This could also be measured for Phase II awards

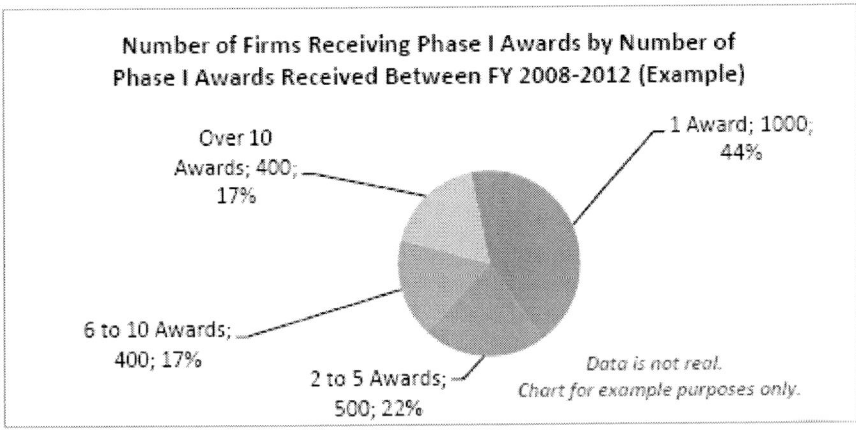

If the chart was based on actual data, one could conclude that over half of the Phase I award winners received multiple awards, with 17% receiving over 10.

2. *Women-owned Small Business Concern (WOSB) Outreach:* • Are more women-owned businesses applying to the program? • Are they receiving awards? • Is the awardor success rate for women-owned firms different than for other firms? Rationale: Congress has identified this group to be of particular interest.	• Number and percent of applications by women-owned firms compared to historic numbers and rates of participation by WOSBs • Number and percent of awards and award dollars to women-owned firms • Award rate: Number of WOSB awards /Number of WOSB applications as compared to the number of awards and number of applications submitted by non-WOSB firms
3. *Socially and Economically Disadvantaged Small Business Concern (SDB) Outreach:*[4]	

[4] 13 CFR section 124 (i.e., 8(a).

• Are more disadvantaged businesses applying to the program? • Are they receiving awards? • Is the awardor success rate for disadvantaged firms different than for other firms? Rationale: Congress has identified this group to be of particular interest.	• Number and percent of applications by disadvantaged firms compared to historic numbers and rates of participation by SDBs • Number and percent of awards and award dollars to disadvantaged firms • Award Rate: Number of SDB awards / number of SDB applications as compared to the number of awards and number of applications submitted by non-SDB awardees
4. *Targeted Geographic Outreach:* SBA has identified targeted geographic areas as those states that are in the bottom half of receiving SBIR/STTR awards. • Are more businesses in targeted geographic areas applying to the program? • Are they receiving awards? • Is the awardor success rate rate for businesses in targeted geographic areas different than for firms not in targeted geographic areas? Rationale: Helps identify whether outreach to targeted geographic areas is improving.	• Number of applications by firms in targeted geographic areas as compared to the number of applications from firms in those geographic areas over the previous 5 fiscal years • Number of awards to firms in targeted geographic areas • Award rate: Number of awards /Number of applications
5. *Outreach Activity Assessment*: • What is the most effective way to increase outreach to the general population?	 • New applicants should identify how they learn about the program.

• What is the most effective way to increase outreach to women-owned small business cocnerns? • What is the most effective way to increase outreach to disadvantaged small business concerns? • What is the most effective way to increase outreach to targeted geographic areas? Rationale: Helps identify whether new initiatives are effective in increasing applicant pool. However, it should be recognized that it may take years before businesses may apply for an award after hearing about the SBIR program.	This should then be analyzed via a bar chart to identify what activities are working for each targeted population. An example is shown below • Cost per applicant can also be assessed based on the cost of the activity in comparison to the number of applicants it attracted or the increase in number

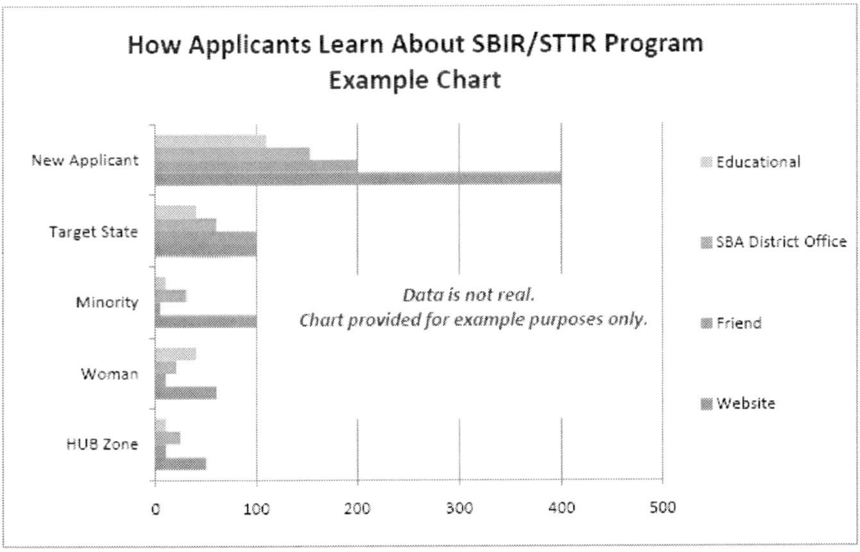

If the above chart contained actual data, one could conclude that the website is the primary source for learning about the program. However, educational events appear to effectively target women-owned firms.

The exact categories in the above chart would depend on the marketing activities we wanted to assess. Applicants should be allowed to choose multiple categories.

b. Compliance and Efficiency

Rationale: These metrics are to ensure compliance with the requirements set forth by statute, the policy directives, and to assess the process.

Assessment Questions	Potential Proposed Metrics
1. *Funding Level:* a. Are participating agencies complying with the minimum level of funding for the SBIR program? (SBIR is 2.6% in FY 2012 and increasing by .1% each year through FY 2017. STTR is 0.35% for FYs 2012 and 2013, with continued increases through 2016.) b. To what extent are agencies not complying with statutory levels?	These should be reviewed on an annual and 3-year basis : • The SBIR and STTR statutory minimum funding levels is determined by multiplying the statutorily prescribed rate by an agency's extramural R/R&D budget • Amount of awards issued by each agency during the previously completed fiscal year • Amount of extramural budget • SBIR and STTR Award Amounts as percentage of extramural budget • Percentage above or below statutory minimum
2. *Processing Times:* a. How long does it typically take to notify applicants in a Phase I solicitation? Phase II solicitation? b. How long does it take to award a Phase I award? Phase II award? c. Are applicants notified whether they were selected for an award within	• Average and/or median days to notify each phase • Average or median days to award each phase • Percentage of awards meeting the statutory timeline requirement for

Assessment Questions	Potential Proposed Metrics
the statutorily required time period? • National Institutes of Health (NIH) or the National Science Foundation (NSF): 1 year; • All other agencies: 90 days after solicitation closes d. Are awards issued within the time period advised by the Policy Directives? • NIH or NSF: 15 months • All other agencies: 180 days	notification of selection as an awardee • Percentage of awards meeting the Policy Directives' timeline guidance for issuance of awards
3. *Data Collection:* a. Are agencies submitting their final reports on time? b. To what extent are agencies overdue in submitting their final report?	 • Percentage of reports received on time • Number of calendar days overdue
4. *Fraud Waste and Abuse:* a. Are agencies reporting incidents of fraud, waste, and abuse (FWA)?	 • Number of FWA incidents reported

c. Commercialization and Other EconomicConsiderations

Assessment Questions / Rationale	Potential Proposed Metrics
1. *Progress to Commercialization:* a. To what extent do Phase I awards transition to Phase II? b. To what extent do Phase I awards achieve some initial level of commercialization (At least $100,000 in sales or investment or received a patent on the technology)? Full commercialization: Sales or investment) at least equal to SBIR/STTR awards?	• Percent of Phase I award recipients (based on number and dollars) that: o Receive a Phase II award o Achieve at least $100,000 in sales or investment or received a patent on the technology o Achieve sales or investment at least equal to the amount of their SBIR/STTR awards

(Continued)

Assessment Questions / Rationale	Potential Proposed Metrics
c. To what extent do Phase II awards achieve some initial level of commercialization (At least $100,000 in sales or investment or received a patent on the technology)? Full commercialization: Sales or investment) at least equal to SBIR/STTR awards?	• Percent of Phase II award recipients (based on number and dollars) that: ○ Achieve at least $100,000 in sales or investment or received a patent on the technology Achievesales or investment at least equal to SBIR/STTR awards

Rationale: If we consider the SBIR/STTR portfolio to be comparable to a large early/seed stage venture capital fund, Phase I would be considered the initial investment and Phase II the follow-on investments. As such, we should expect that a large percentage of investments will fail, some investments will have moderate returns, and some will be home runs. Understanding the overall success of the SBIR/STTR portfolio is critical to evaluating this program.

Results for seed and early stage investments may take years to realize, especially for life sciences. One way to track this progression would be to measure commercialization results by "vintage year" or the year the initial (Phase I) investment was made. An example of how this metric might be captured is shown below.

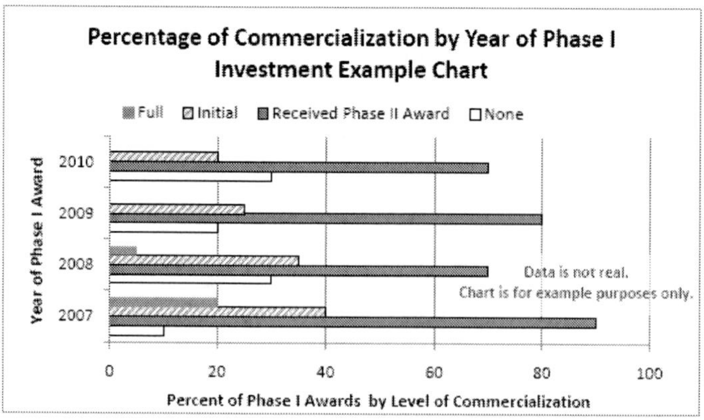

If the chart was based on actual data, one could see increased progress in achieving commercialization as each "vintage year" of awards matured. This would give the IPC and Congress a sense for both the eventual percentage of realizations and how long it takes before one can reasonably expect commercialization results. Once sufficient data was gathered, one can calculate an average commercialization growth rate from one year of Phase I investment.

Similar tracking could also be done for Phase II investments. However, one would expect a much higher percentage of Phase II investments to achieve commercialization.

The percentage of commercialization that should be achieved is debatable. Too high a number may indicate too conservative an approach which does not yield the "home run" returns associated with winners in an early stage portfolio. As a benchmark, we could look at successful early stage funds.

SBA's Participating Security Small Business Investment Companies (SBICs) may provide a reasonable indicator since they performed a high percentage of seed and early stage investing. The chart below distributes investment return data for successful Participating Securities SBICs.

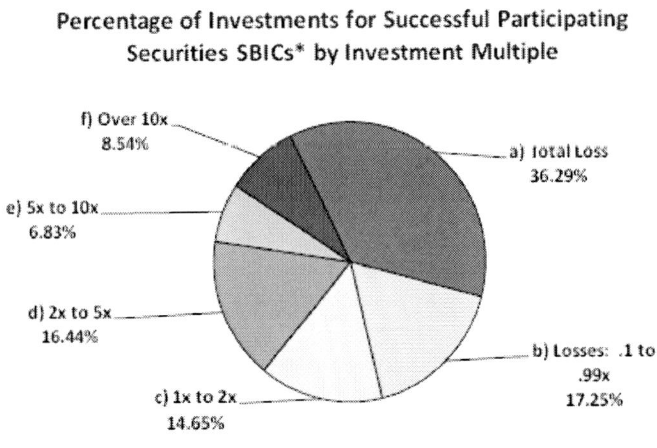

*Notes:

- Chart is based on SBA SBIC realization data, collected since 1997.
- Only successful Participating Securities SBICs licensed between 1997 and 2004 were included in this data set. A successful fund is defined as returning at least 70% of all capital paid into the funds (on an unleveraged basis) with sufficient value remaining to return all capital. This included 72 SBICs with 1,229 portfolio companies.
- Investment Multiple = Total Proceeds/Total Investment Cost.

As shown, over half of the investments were losses, with a small portion hitting home runs. It should also be noted that less than 15% of these SBICs investments produced over half of the returns. One might expect similar or higher risk results in the SBIR/STTR portfolio since SBIR/STTR targets high-risk, early-stage work with potential for technological innovation.

Alternatively, the IPC could hire a consultant to establish reasonable benchmarks.

Assessment Questions / Rationale	Proposed Metrics
2. *Return on Investment:* a. What is the "return on investment" for SBIR and STTR investments? b. Do venture backed companies have a higher "return on investment" than SBIR and STTR investments? Rationale: Although SBIR and STTR "returns" are difficult to measure, sales, number of patents filed and/or received,and investments since the awards can provide a useful measurement. It is important to try to capture this information in order to assess the effectiveness of the program. This may also provide an indicator of how venture backed companies perform relative to non-venture backed companies.	• Return on Investment = (Sales and investments since the Phase II award – SBIR Award Amounts)/SBIR Award Amounts • Compare venture backed companies to non-venture backed companies • Similar to the commercialization benchmarks this may be best captured by the "vintage year" of the Phase I or Phase II investment
3. *Jobs:* a. How many jobs are created or sustained by the technology? Rationale: As a Federal program designed to improve the economy, jobs are a relevant result. However, jobs are historically difficult to track and this metric may need to be reassessed based on the availability of the data.	• The number of jobs associated with the technology should be assessedto determine the number of jobs sustained and created

INDEX

A

B

C

Q

R

S

T

U

V

W

Y